T0200080

The SCIENCE of PLANTS
INSIDE THEIR SECRET WORLD

The SCIENCE of
PLANTS
INSIDE THEIR SECRET WORLD

Royal Botanic Gardens
KEW

Penguin
Random
House

DK LONDON

Senior Editor Helen Fewster
Project Editors Nathan Joyce,
Anna Limmerick, Simon Maughan,
Steve Setford, David Summers,
Angela Wilkes, Miezan van Zyl
Editorial Assistant Daniel Byrne
DK Media Archive Romaine Werblow
Production Editor Jacqueline Street-Elkayam
Managing Editor Angeles Gavira Guerrero
Associate Publishing Director Liz Wheeler
Publishing Director Jonathan Metcalf

Senior Art Editor Sharon Spencer
Project Art Editors Ray Bryant,
Phil Gamble, Vanessa Hamilton
Photographer Gary Ombler
Illustrators Dominic Clifford, Alex Lloyd
Jacket Designer Akiko Kato
Jacket Design Development Manager Sophia MTT
Senior Production Controller Meskerem Berhane
Managing Art Editor Michael Duffy
Art Director Karen Self
Design Director Phil Ormerod

DK DELHI

Senior Picture Researcher
Surya Sankash Sarangi
Senior Jackets Coordinator
Priyanka Sharma-Saddi
Managing Jackets Editor Saloni Singh

Picture Research Manager Taiyaba Khatoon
Senior Jacket Designer Suhita Dharamjit
DTP Designer Rakesh Kumar
Senior DTP Designer Harish Aggarwal
Production Manager Pankaj Sharma

Content previously published as *Flora* in 2018
This edition published in Great Britain in 2022 by
Dorling Kindersley Limited
DK, One Embassy Gardens, 8 Viaduct Gardens,
London, SW11 7BW

The authorised representative in the EEA is
Dorling Kindersley Verlag GmbH. Arnulfstr. 124,
80636 Munich, Germany

Copyright © 2018, 2022 Dorling Kindersley Limited
A Penguin Random House Company
10 9 8 7 6 5 4 3 2 1
001–326348–May/2022

All rights reserved. No part of this publication may be reproduced, stored in
or introduced into a retrieval system, or transmitted, in any form, or by any
means (electronic, mechanical, photocopying, recording, or otherwise),
without the prior written permission of the copyright owner.

A CIP catalogue record for this book
is available from the British Library.
ISBN: 978-0-2415-1550-1

Printed in China

For the curious
www.dk.com

Contributors

Jamie Ambrose was an author, editor, and Fulbright scholar with a special interest in natural history. Her books include DK's *Wildlife of the World*.

Dr Ross Bayton is a botanist, taxonomist, and gardener with a passion for the plant world. He has authored books, magazine features, and scientific papers encouraging readers to understand and appreciate the importance of plants.

Matt Candeias is the author and host of the *In Defense of Plants* blog and podcast (www.indefenseofplants.com). Trained as an ecologist, Matt focuses most of his research on plant conservation. He is also an avid gardener, both indoors and out.

Dr Sarah Jose is a professional science writer and language editor with a PhD in botany, and a deep love of plants.

Andrew Mikolajski is the author of over 30 books on plants and gardening. He has lectured in garden history at the English Gardening School at Chelsea Physic Garden and for the Historic Houses Association.

Esther Ripley is a former managing editor who writes on a range of cultural subjects, including art and literature.

David Summers is a writer and editor with a training in natural history filmmaking. He has contributed to books on a range of subjects including natural history, geography, and science.

The Royal Botanic Gardens, Kew

The Royal Botanic Gardens, Kew is a world famous scientific organization, internationally respected for its outstanding collections as well as its scientific expertise in plant diversity, conservation, and sustainable development around the world. Kew's 132 hectares (320 acres) of landscaped gardens, and Wakehurst – Kew's Wild Botanic Garden in Sussex – attract over 2.5 million visits every year. Kew was made a UNESCO World Heritage Site in July 2003 and celebrated its 260th anniversary in 2019. Wakehurst is home to Kew's Millennium Seed Bank, the largest wild plant seed bank in the world. RBG Kew receives approximately one third of its funding from Government through the Department for the Environment, Food and Rural Affairs (Defra) and research councils. Further funding needed to support RBG Kew's vital work comes from donors, membership and commercial activities including book sales.

Half-title page Bird of paradise (*Strelitzia reginae*)
Title page Northern sea oat (*Chasmanthium latifolium*)
Above Autumn trees changing colour
Contents page *Nigella papillosa* 'African Bride'

MIX
Paper from
responsible sources
FSC™ C018179
www.fsc.org

This book was made with Forest Stewardship Council™ certified paper – one small step in DK's commitment to a sustainable future. For more information go to www.dk.com/our-green-pledge

contents

the plant kingdom

12 what is a plant?
14 types of plant
16 classifying plants

roots

20 fibrous roots
22 tap roots
24 plant supports
26 absorbing nutrients
28 storage systems
30 plants in art:
     impressions of nature
32 nitrogen fixing
34 the fly agaric
36 contractile roots
38 aerial roots
40 the strangler fig
42 living on air
44 parasitic plants
46 mistletoe
48 roots in water
50 water lilies
52 breathing roots
54 mangroves

stems and branches

58 types of stem
60 inside stems
62 plants in art:
     natural renaissance
64 tree trunks
66 types of bark
68 the quaking aspen
70 branch arrangement
72 winter buds
74 insulated buds
76 cauliflorous plants
78 stem defences
80 the kapok tree
82 stems and resin
84 the dragon tree
86 storing food
88 beneficial relationships
90 fibrous trunks
92 plants in art:
     west meets east
94 twining stems
96 climbing techniques
98 storage stems
100 the saguaro cactus
102 leafless stems
104 new plants from stems
106 moso bamboo

leaves

110 types of leaf
112 leaf structure
114 compound leaves
116 plants in art:
       designed by nature
118 developing leaves
120 unfurling ferns
122 leaf arrangement
124 plants in art:
       botanic reinvention
126 leaves and the water cycle
128 leaves and light
130 leaf size
132 leaf shapes
134 drip tips
136 leaf margins
138 hairy foliage
140 plants in art:
       ancient herbals
142 succulent foliage
144 waxy leaves
146 the quiver tree
148 silver leaves
150 variegated leaves
152 the japanese maple
154 autumn colours
156 stinging leaves
158 the sweet thorn
160 leaf defences
162 leaves in extremes
164 the bismarck palm
166 floating leaves
168 plants with pools
170 leaves that eat
172 sundews
174 leaves that reproduce
176 protective bracts
178 types of bract
180 leaves and spines

flowers

184 parts of a flower
186 ancient flowers
188 flower shapes
190 flower development
192 flowers and seasons
194 plants in art:
       chinese painting
196 flower fertilization
198 pollen grains
200 encouraging diversity
202 managing pollen
204 plants in art:
       japanese woodblock prints
206 unisexual plants
208 incompatible flowers
210 the titan arum
212 flowers that change sex
214 inflorescences
216 types of inflorescence
218 rays and discs
220 the sunflower
222 wind-pollinated flowers
224 grass flowers
226 flowers and nectar
228 storing nectar
230 designed for visitors
232 precision pollination
234 buzz pollination
236 plants in art:
       radical visions
238 winter blooms
240 flowers for birds
242 flowers for animals
244 flying visitors
246 the queen of the night
248 colour attraction
250 nectar guides
252 the hollyhock
254 colour signals

256 restricted entry
258 fragrant traps
260 the corpse flower
262 special relationships
264 natural mimics
266 designed to deceive
268 self-pollinating flowers
270 plants in art:
       blooms for royalty
272 closing at night
274 bud defences
276 armoured flowers
278 colourful bracts
280 plants in art:
       american enthusiasts
282 cone reproduction

seeds and fruits

286 seed structure
288 naked seeds
290 inside seed cones
292 enclosed seeds
294 types of fruit
296 plants in art:
       ancient gardens
298 flower to fruit
300 fruit anatomy
302 the banana
304 seed distribution
306 dispersed by diet
308 colourful seeds
310 plants in art:
       art and science
312 animal couriers
314 seeds with wings
316 the dandelion
318 seeds with parachutes
320 silky seeds
322 common milkweed
324 pods and capsules
326 exploding seedpods
328 nigella
330 seeds and fire
332 water dispersal
334 plants in art:
       painting the world
336 natural clones
338 fern spores
340 spore to fern
342 haircap mosses

344 glossary
350 index
358 list of botanical art
359 acknowledgments

BLUE LOTUS (*NYMPHAEA CAERULEA*)

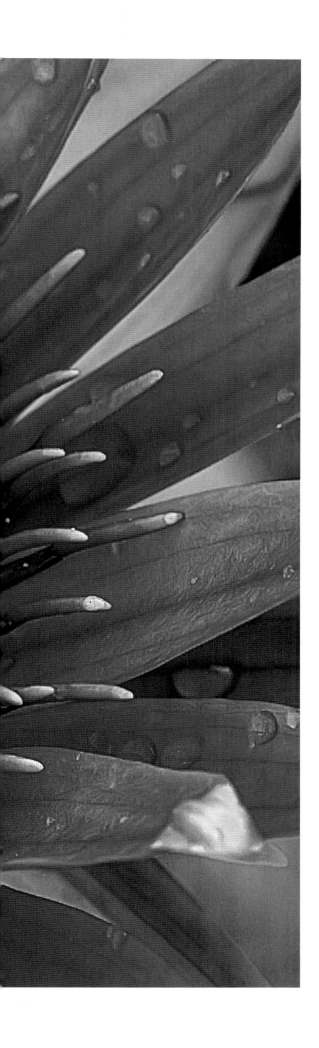

foreword

66 Many plants of many kinds ... from so simple a beginning endless forms most beautiful and most wonderful have been, and are being, evolved. 99

So ends the book that transformed our world view like no other – *On the Origin of Species* by Charles Darwin. More than 160 years after its publication, we are still at the dawn of understanding just how many plant forms there are, and their precise role in nature. During this period, the number of vascular plant species known to science has grown from tens of thousands to a global estimate of 345,000 species, with some 2,000 new additions every year.

This book offers an excellently written, and superbly illustrated, framework for peering into the complex lives of plants. Just as physicians need a system to understand the body, so plant lovers profit from a standard set of names and concepts for communicating and learning effectively about their subject.

The advantages of studying plants go far beyond the realm of fundamental knowledge. Plants have long provided us with invaluable assets, from food to fibre, shelter, and medicines. They have coped with, and often overcome, similar challenges to those our own species increasingly faces: heat, drought, nutrient shortage, and even viruses and bacteria. Understanding how plants – through the evolutionary processes first described by Darwin – have solved problems thrown at them by their environment, and applying these solutions to our own challenges, is now a promising area of research, biomimicry.

However, amid the anticipation of scientific discoveries yet to be made is the realization that plant diversity has never been more at risk. Scientists now estimate that 2 in 5 plant species are threatened with extinction. Left to their fate, we will not only lose the products and services they provide, but also a dangerously large proportion of the functions that sustain healthy ecosystems.

By reading this book, I hope you join Darwin, myself, and so many others in developing a deep fascination for plants – and that it inspires you to join global efforts to protect and spread appreciation for their wonderful world.

PROFESSOR ALEXANDRE ANTONELLI
DIRECTOR OF SCIENCE
ROYAL BOTANIC GARDENS, KEW

the plant kingdom

plant. a living organism, usually containing chlorophyll – including trees, shrubs, herbs, grasses, ferns, and mosses – that typically grows in a permanent position, absorbing water and inorganic substances through its roots, and synthesizing nutrients in its leaves by photosynthesis.

WHAT IS NOT A PLANT?

Although they may look like plants, fungi are actually more closely related to animals. They cannot produce their own food – as plants do – and rely on the carbohydrates formed by normal plants. Many modern plants, notably forest trees and orchids, maintain some degree of dependence on fungi (see p.34). Algae is a general term used to denote a diverse range of organisms, including seaweeds, that do not have true roots, stems, or leaves, though they may appear green and leafy. Most of them live in water, and they are dominant in the sea. Lichens are composite organisms formed from algae (or certain bacteria) and fungi that have a mutually beneficial (symbiotic) relationship.

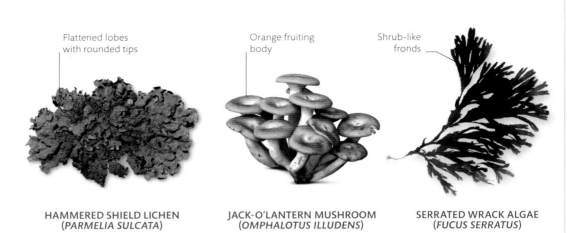

Flattened lobes with rounded tips

Orange fruiting body

Shrub-like fronds

HAMMERED SHIELD LICHEN
(*PARMELIA SULCATA*)

JACK-O'LANTERN MUSHROOM
(*OMPHALOTUS ILLUDENS*)

SERRATED WRACK ALGAE
(*FUCUS SERRATUS*)

what is **a plant?**

Plants are living organisms found virtually all over the planet, other than in permanently frozen or totally arid places. They range in size from majestic trees to tiny plants no larger than a grain of rice. Originally, all plants were aquatic and had roots simply to anchor them in place. Once they moved on to land, many formed an association with fungi, which helped their roots obtain water and minerals. Plants differ from other organisms because they can make their own food by photosynthesis. Using chlorophyll – a green pigment in its cells – a plant absorbs energy from sunlight and uses carbon dioxide in the atmosphere to produce sugars. Unlike animals, which often stop growing once mature, plants may continue to grow and produce new material every year, either to increase their size or to replace lost or damaged material.

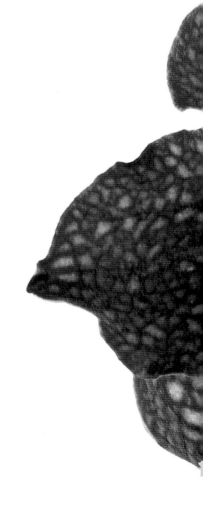

Bright blooms
There are more than 350,000 species of flowering plants. Flowers are not just for show – they are the reproductive structures of a plant, and their shape and colour are there to attract pollinators. These exotic orchid flowers are from a *Vanda* hybrid, many of which are native to tropical regions of Asia.

Opening flower bud

Flowering stems
sometimes
branch

kingdom *Plantae*

non-flowering plants

This includes plants such as ferns, mosses, and liverworts, all of which reproduce from spores. It also encompasses non-flowering plants that produce seeds, such as the gymnosperms, of which conifers are the largest group. Gymnosperms produce naked (not enclosed) seeds.

liverworts	mosses	hornworts	lycopods

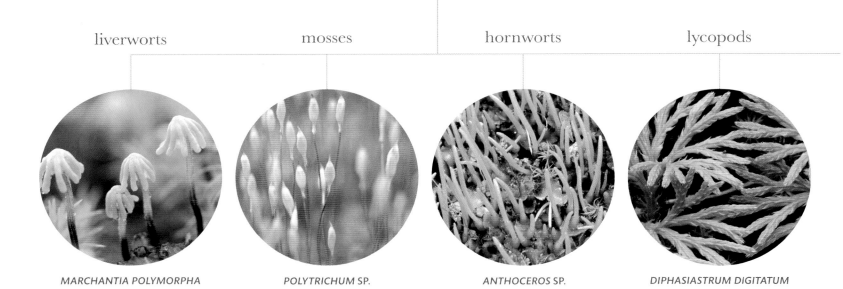

MARCHANTIA POLYMORPHA　　*POLYTRICHUM SP.*　　*ANTHOCEROS SP.*　　*DIPHASIASTRUM DIGITATUM*

types of plant

The smallest, simplest plants are known collectively as bryophytes and include liverworts, mosses, and hornworts. They often grow in places that are permanently damp, such as bogs or on the shady side of rocks and tree trunks. Ferns are an ancient and diverse group of plants that have adapted to a wide range of habitats. They reproduce by means of spores. Gymnosperms produce cones, the females of which bear naked (not enclosed) seeds. Flowering plants are the most diverse and complex type of plant. They bear flowers, and the seeds that they produce are enclosed within fruits.

Evolving complexity
The first plants to grow on land were the ancestors of simple mosses, liverworts, and hornworts. Over the millennia, evolution has led to greater complexity, and angiosperms (flowering plants) now dominate the plant kingdom. From the fossil record it is clear that gymnosperms were once a much larger and more diverse group than they are today.

flowering plants

The angiosperms (flowering plants) are a diverse group and are found in a wide range of habitats around the world. Like gymnosperms, they produce pollen and seeds, but the seeds of an angiosperm are enclosed within a fruit.

ferns

gymnosperms

angiosperms

CYATHEA SP.

LARIX SP.

NYMPHAEA 'MASANIELLO'

magnoliids

monocots

eudicots

MAGNOLIA GRANDIFLORA

LILIUM AURATUM

ROSA RUBIGINOSA

classifying plants

Individual plants are formally named according to the system devised by Swedish botanist Carl Linnaeus (1707–78), which uses two-word names. In Latin and italics, a plant's name is made up of the genus to which it belongs, followed by the name of its particular species. Plants are grouped together according to the characteristics that they share. Historically, classification was determined by a plant's physical characteristics (notably flower structure) and biochemistry (the chemical compounds within a plant), and was often speculative or subjective. Nowadays, genetic evidence provides a more reliable means of understanding the relationships between plants.

A system of classification
This page of illustrations by Georg Dionysius Ehret, published in 1736, illustrates the Linnaean classification system. Linnaeus differentiated species by looking at the sexual parts of the plants, notably the different numbers of male and female parts.

LOTUS FLOWER
(*NELUMBO* SP.)

WATER LILY FLOWER
(*NYMPHAEA* SP.)

PLANE TREE FLOWER
(*PLATANUS* SP.)

Surprising relationships
DNA profiling has led to some unexpected discoveries. People often confuse the lotus (*Nelumbo*) and the water lily (*Nymphaea*), because they look alike. Genetic profiling has now shown, however, that the lotus is in fact more closely related to the plane tree (*Platanus*).

HIERARCHY OF TERMS

Botanists use the following ranks to arrange plants. Plants are organized into divisions, then classes and so on, into increasingly specific groups, according to the structure of their flowers, fruits, and other parts, as well as evidence from fossils and DNA profiling.

division
separates plants according to key features, e.g. angiosperms and gymnosperms

class
divides plants according to fundamental differences, such as monocots and eudicots

order
groups families with a common ancestor

family
contains plants that are clearly related, e.g. the rose family

genus (genera)
a group of closely related species with similar features

species
plants with shared characteristics, which can sometimes interbreed

subspecies, forms, varieties
have features that differ from those typical of a species, or that are geographically distinct

cultivar
a cultivated variety of a plant that was produced from natural species or hybrids

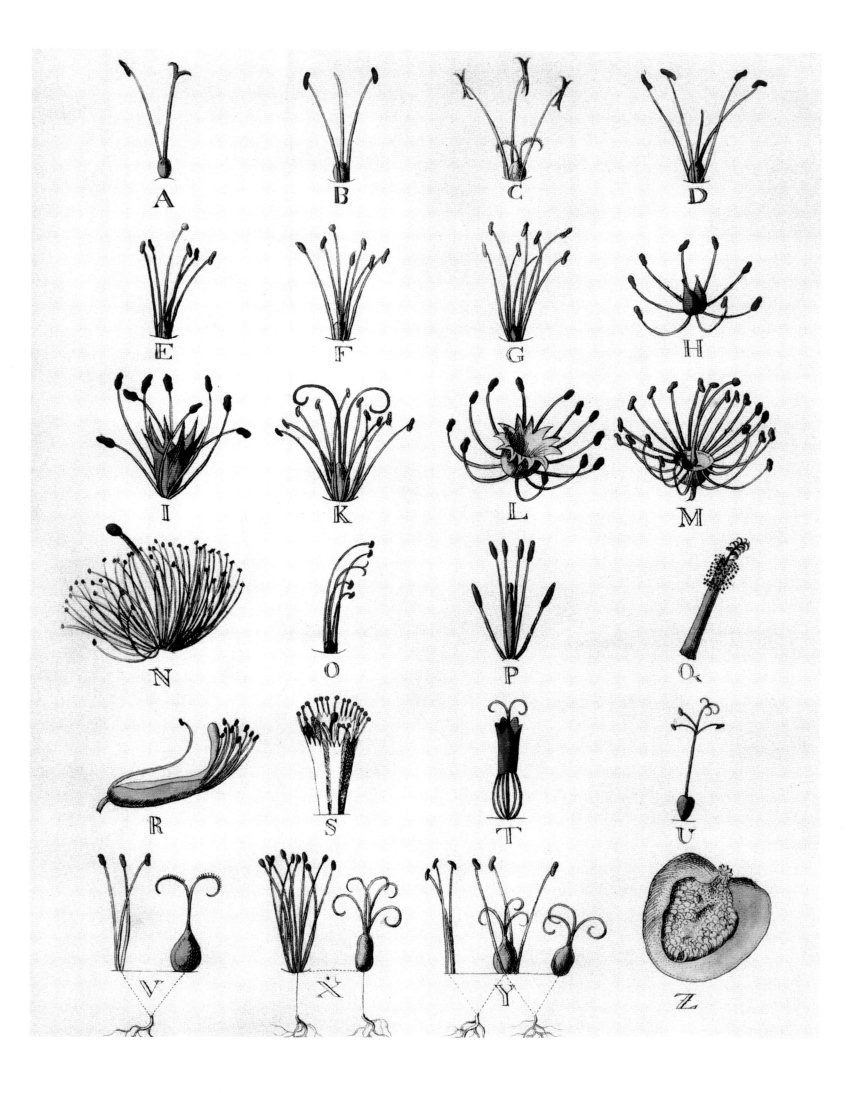

roots

root. the part of a plant, usually underground, that anchors it in the earth, and transports water and nutrients to the rest of the plant.

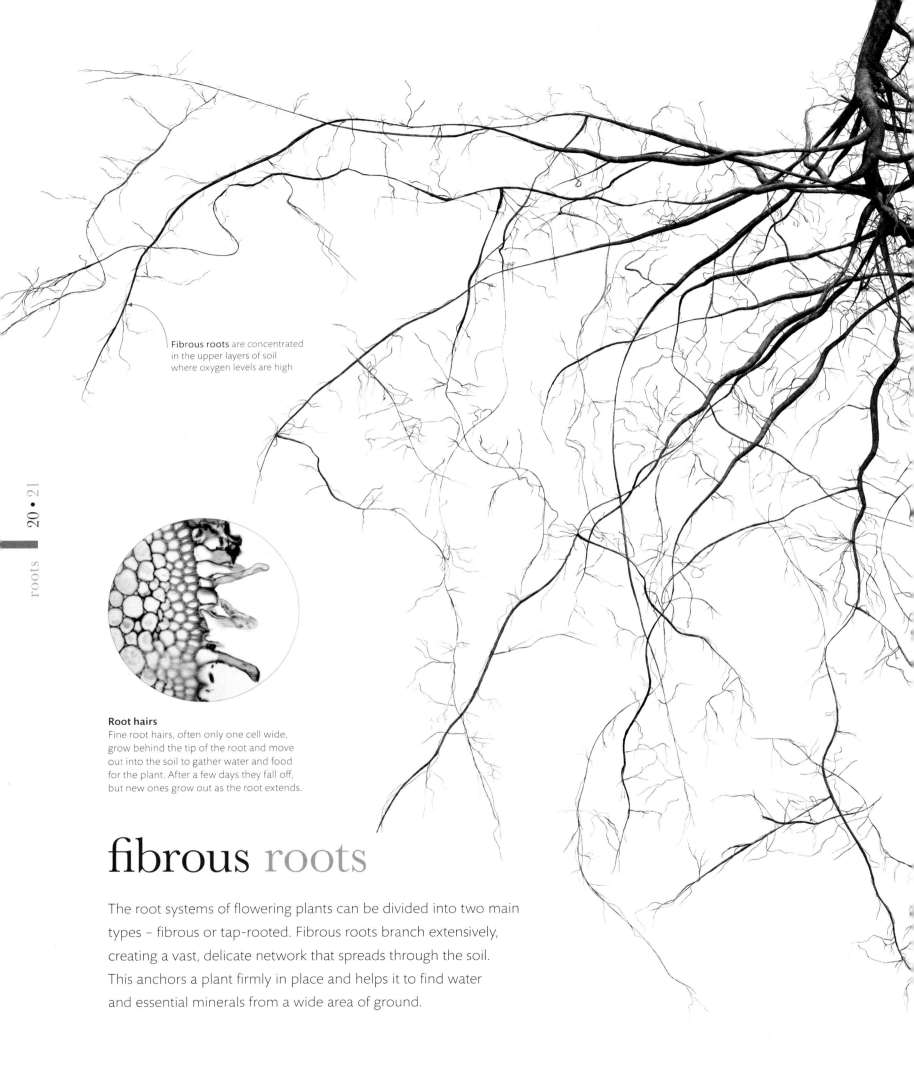

Fibrous roots are concentrated in the upper layers of soil where oxygen levels are high

Root hairs
Fine root hairs, often only one cell wide, grow behind the tip of the root and move out into the soil to gather water and food for the plant. After a few days they fall off, but new ones grow out as the root extends.

fibrous roots

The root systems of flowering plants can be divided into two main types – fibrous or tap-rooted. Fibrous roots branch extensively, creating a vast, delicate network that spreads through the soil. This anchors a plant firmly in place and helps it to find water and essential minerals from a wide area of ground.

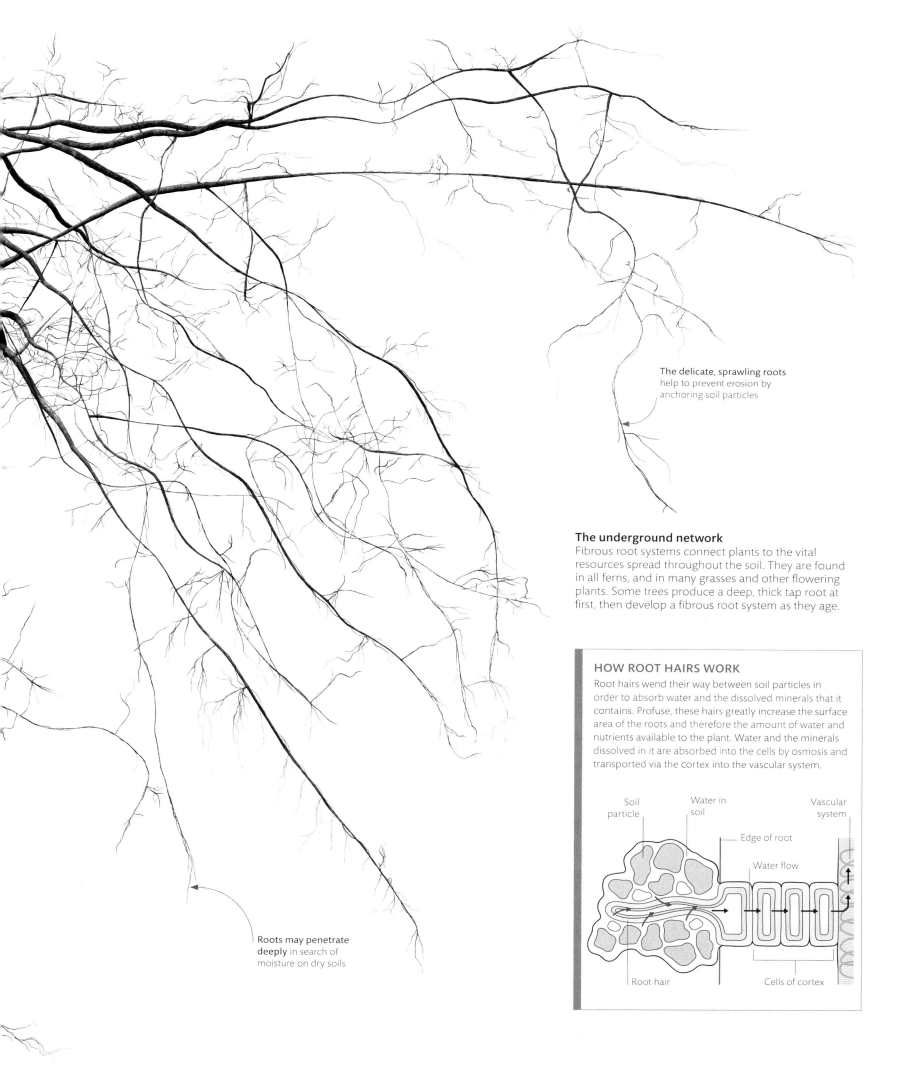

The delicate, sprawling roots help to prevent erosion by anchoring soil particles

The underground network
Fibrous root systems connect plants to the vital resources spread throughout the soil. They are found in all ferns, and in many grasses and other flowering plants. Some trees produce a deep, thick tap root at first, then develop a fibrous root system as they age.

HOW ROOT HAIRS WORK
Root hairs wend their way between soil particles in order to absorb water and the dissolved minerals that it contains. Profuse, these hairs greatly increase the surface area of the roots and therefore the amount of water and nutrients available to the plant. Water and the minerals dissolved in it are absorbed into the cells by osmosis and transported via the cortex into the vascular system.

Soil particle

Water in soil

Vascular system

Edge of root

Water flow

Root hair

Cells of cortex

Roots may penetrate **deeply** in search of moisture on dry soils

The tap root swells as it grows, storing food in the form of carbohydrates

Biennial tap roots, like beetroot, grow in year one, then die in year two when the plant flowers and sets seed

tap roots

In contrast to fibrous root systems, tap-rooted plants normally develop a single dominant root with much smaller side roots. Some trees can develop either type – a tap root on loose soils, but a fibrous network on heavy soils. Most seedlings in the eudicot group of flowering plants (see p.15) start life tap-rooted, with the seed root – or radicle – dominating the root system. If the plant is not genuinely tap-rooted, this eventually dies off, causing the roots to branch into a fibrous arrangement.

The end of the tap root is long and slender so that it can penetrate deep into the soil

The deep red colour comes from betalain pigments, which can be used as dyes or food colouring

Tap root food crops
Some plants, such as beetroot (*Beta vulgaris*), store carbohydrates as sugars within their tap roots. These energy reserves are used to produce flowers and seeds (see pp.28–29). Tap root crops must be harvested while the roots are still rich in sugars, before the plant is able to bloom; after flowering, the roots become woody and unpalatable.

Fuelled by sugars, the tap root can produce new leaves and stems to replace any harvested as salad crops

Thin side roots take up most of the water through minute root hairs

DEEP ROOTS

Tap roots bring several advantages to their owners. Often growing deep into the soil, they can access water and nutrients beyond the reach of shallow-rooted plants. For weeds such as dandelions (*Taraxacum officinale*), a tap root also makes it difficult to remove the plant from the soil. More often than not, the leaves break off, leaving the intact root in place and ready to grow back. Many tap roots can regrow from fragments left in the soil; even the smallest piece of dandelion root readily re-sprouts.

TARAXACUM OFFICINALE

Léontodon Taraxacum.

plant supports

Anchorage is one of the major functions of any root system. For most plants, this is accomplished entirely underground, but where the soil is shallow – as in many rainforests – some species produce elaborate above-ground support systems. Coastal mangroves, growing in loose and unstable soil, do the same (see pp.54–55). Buttress, prop, and stilt roots all help to hold aloft the largest and heaviest of canopies, assuring a solid foundation.

SUPPORTIVE ROOT SYSTEMS

Buttress roots develop on shallow soils and are part of the primary root system. In contrast, prop and stilt roots develop from the main stem and branches above. Prop roots – common in corn plants – brace slender trunks and often appear in tiers. Stilt roots drop down from side branches.

Roots flare out to brace trunk

Buttresses direct water to base of tree

BUTTRESS

Tall, slender stems benefit from extra support

Props are aerial roots, growing from the stem

New prop roots form as the tree gets taller

PROP

Stilts hold the tree firm against tidal flows

Side branches develop into stilt roots

STILT

Buttress roots

Coastal trees such as the endangered looking-glass mangrove (*Heritiera littoralis*) need strong supports to protect against tidal flooding. Buttress roots interweave for greater strength and, if the tree's crown is larger on one side, more buttresses develop on the opposite side for stability.

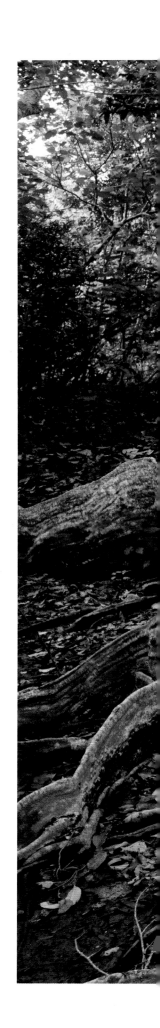

Hydrangeas on acid soils
Most hydrangeas have white flowers, but the colour of mophead hydrangeas (*Hydrangea macrophylla*) is determined by the acidity of the soil. On alkaline soils above pH7, they are normally red to pink. In acid soils below pH7, aluminium becomes soluble, and is absorbed by hydrangea roots. The aluminium ions bind to the red pigments in the blooms, and turn the flowers blue.

Pink mopheads
occur on hydrangeas in alkaline soils, where aluminium is unavailable

Healthy green leaves
depend on the roots absorbing sufficient nutrients, such as magnesium and iron

A blue mophead
develops when the hydrangea grows in acid soil, where aluminium can be absorbed

absorbing nutrients

When roots absorb moisture from the soil, they also take up vital minerals dissolved in the water that a plant needs in order to grow and thrive. The chemical structure of soil varies widely from one place to another, and deficiencies in the key elements that a plant requires can lead to stunted growth or discoloured leaves. Leaves are, however, able to store surplus nutrients for later use, when there might be a shortage.

Iron deficiency
Plants need iron to manufacture important enzymes and pigments, including chlorophyll, the light-sensitive green pigment necessary for photosynthesis. If iron is in short supply, a plant produces less chlorophyll and its leaves turn yellow, as in this hydrangea. Iron can only be absorbed by the roots when dissolved in water, so on alkaline soils, where iron is less soluble, plants are more likely to suffer from iron deficiency.

FOOD DISTRIBUTION
Nutrients absorbed from the soil are carried up from the roots to every part of the plant – stems, branches, leaves, and flowers – through bundles of miniscule pipes (xylem). The three most important minerals for a plant are nitrogen, phosphorus, and potassium. These are also the main constituents of many garden fertilizers.

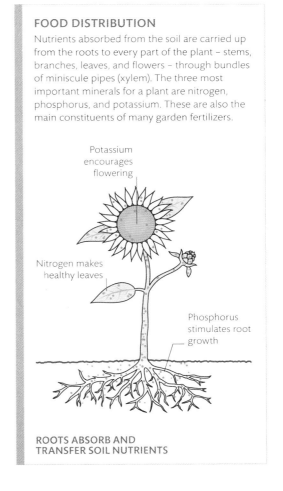

Potassium encourages flowering

Nitrogen makes healthy leaves

Phosphorus stimulates root growth

ROOTS ABSORB AND TRANSFER SOIL NUTRIENTS

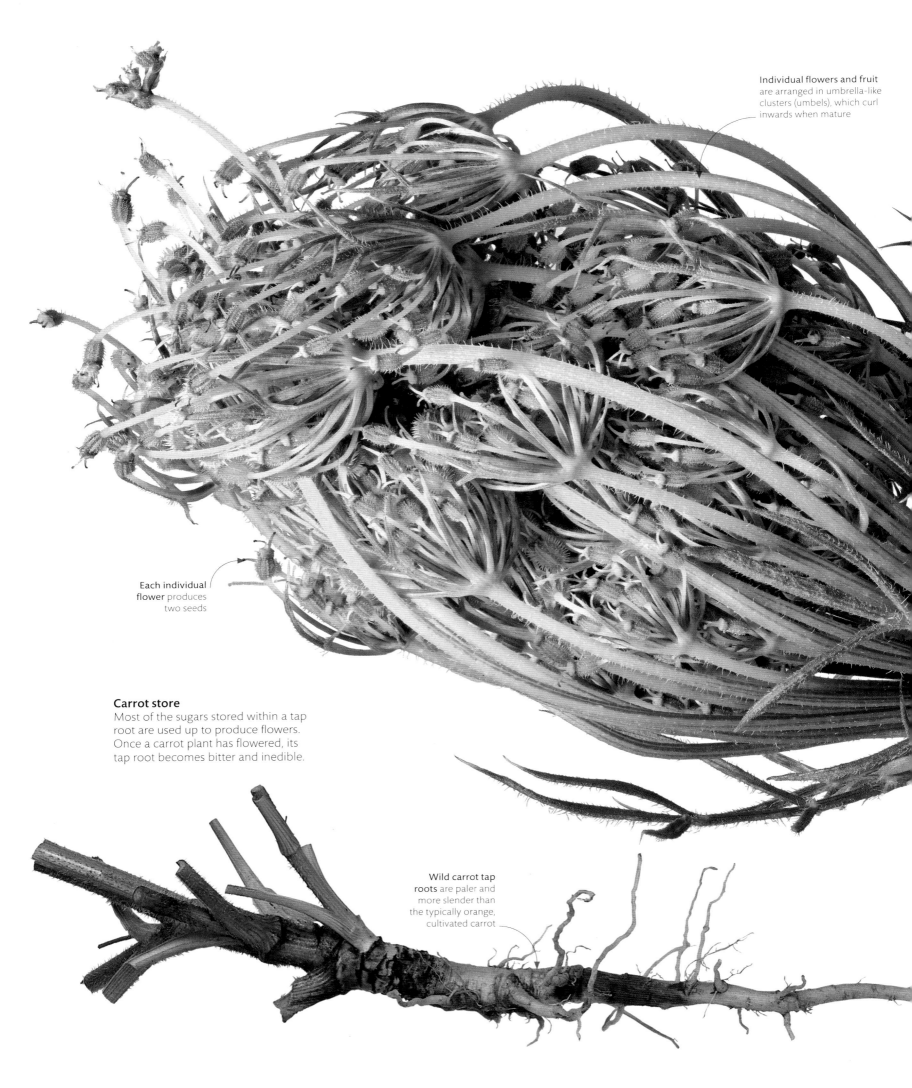

Individual flowers and fruit are arranged in umbrella-like clusters (umbels), which curl inwards when mature

Each individual flower produces two seeds

Carrot store

Most of the sugars stored within a tap root are used up to produce flowers. Once a carrot plant has flowered, its tap root becomes bitter and inedible.

Wild carrot tap roots are paler and more slender than the typically orange, cultivated carrot

Branching, hairy bracts extend under the carrot flower heads and may help to distribute seeds in the wind

UNDERGROUND ENERGY STORAGE

Carrots (*Daucus carota*) originate from Europe and Southwest Asia. The wild plants have slender white roots, but years of cultivation by humans has resulted in the domesticated carrot, which has a much larger and more colourful root. Carrots live for two years. In the first year they develop a crown of leaves and a swelling carbohydrate-packed root. These food reserves are used to make flowers in the second year.

CARROT LIFE CYCLE

Carrot seedlings produce two seed leaves

Flowers develop and exhaust the root's reserves

A tap root develops under new leaves

Food produced by the leaves is stored in the root

At the beginning of year two, the root begins to release its reserves

storage systems

Reproduction takes a lot of energy. Producing flowers, sugary nectar, and seeds requires a substantial investment from a plant. Biennials – plants that live for just two years – allow themselves a whole year in which to build up a reserve of carbohydrates before flowering in their second year of life. Once seeds are set, the exhausted plants die. Many biennial root crops store their reserves in tap roots, but farmers interfere and harvest the nutrient-rich roots before the plants can use their resources to bloom.

The thickened flower stem transports carbohydrates away from the tap root

Carrot tap roots are thick, long, and usually unbranched

Food for seeds

Once the tap root has released the energy reserves stored within and the carrot has flowered, seeds develop inside the curled inflorescence. This can detach, scattering the seeds as it blows around in the wind. At this point, the tap root is shrivelled and exhausted from the effort of producing hundreds of seeds.

impressions of nature

Reacting against the formal constraints and rules of academic painting, the 19th-century French Impressionist painters immersed themselves in the natural world and worked at easels outdoors, so that they could capture their fleeting impressions of the changing light on the landscape in a fresh and spontaneous way. The Postimpressionists who followed them took this approach a step further. Inspired by the geometric forms and dazzling colours that they saw in the natural world, they created expressive, vibrant, semi-abstract works bursting with energy.

Inspired by the Impressionists' fresh new approach to art, Postimpressionists such as van Gogh and Cézanne adopted a simplified approach to painting, with little attempt at realism, and developed a visual language of their own. They emphasized the bold lines, geometric shapes, and luminous colours of the southern French landscape, paving the way for the 20th-century abstract artists to follow.

Vincent van Gogh painted from nature, but he used intense colour and powerful brushstrokes to express his emotional response to it. In his search for a modern approach to art, he was greatly influenced by Japanese prints, with their bold, delineated shapes and flat expanses of colour. His letters from the South of France simmer with joy. Believing he had found "his Japan", he wrote ecstatically about colour and shape. "A meadow full of very yellow buttercups, a ditch with iris plants with green leaves… a strip of blue sky." He was preoccupied by the cypress trees and astonished that no one had painted them as he saw them.

Tree Roots, 1890
This oil on canvas painting by Vincent van Gogh looks at first like a jumble of bright colours and abstract forms. It is in fact a cropped study of gnarled tree roots, trunks, and boughs growing on a quarry slope. The colours are patently unreal. It was painted the morning before van Gogh died and is not finished, but it is astonishing for its vigour: it is full of sun and life.

66 I'll always love nature here, it's something like Japanese art, once you love that you don't have second thoughts about it. 99

VINCENT VAN GOGH, *LETTER TO THEO VAN GOGH*, 1888

Japanese influence
Cypress trees, a polychrome and gold-leaf screen, was painted *c.*1590 by Kanō Eitoku, a major artist of the Kanō school of Japanese painting. The cropped composition, swirling, expressive use of line, and the flat expanses of strong colour are typical of the Japanese style of art that influenced van Gogh.

Clover's three-leaflet
foliage is reflected in
its Latin name, *Trifolium*

Clover stems and
foliage provide
useful protein for
grazing animals

Enriching the soil
Red clover (*Trifolium pratense*) is grown by farmers
as a cover crop, in order to protect bare soil in winter
from soil erosion and nutrient depletion. Because it is
able to harness nitrogen, clover grows vigorously. In
spring, it is ploughed back into the soil, enriching it
with nitrogen. These benefical properties make it
ideal in crop-rotation systems.

Clover flowers are an important source of nectar for bees and other insects

Nitrogen-fixing legumes
Many different plants can fix atmospheric nitrogen into a form they can absorb, including sea buckthorn (*Hippophae*), California lilac (*Ceanothus*), and alder (*Alnus*). It is most common, however, in the legume family, which includes peas, beans, and trefoils, such as this *Trifolium subterraneum* (right). Legume nodules host several different sorts of bacteria, collectively known as rhizobia, that are essential for fixing nitrogen.

Fresh green leaves
have a ready supply of nitrogen; lack of nitrogen can cause yellowing

nitrogen fixing

As a major component of proteins – crucial building blocks of life – nitrogen is an essential element for plants. While it is abundant in the air, it is largely unreactive and unusable, so plants make use of nitrogen-bearing compounds in the soil and absorb them through their roots. Some plants, however, are able to absorb atmospheric nitrogen in cooperation with bacteria and convert it into usable compounds. They are known as nitrogen-fixers.

ROOT NODULES

Plants are only able to fix nitrogen because of the symbiotic relationship between bacteria and the host plant. This takes place in specially modified structures on the roots called nodules, which form when bacteria invade the cortex (outer layer) of growing root hairs. Within the nodules, bacteria produce an enzyme called nitrogenase. This converts gaseous nitrogen into soluble ammonia, a form in which the nitrogen can be used by the plant. In return, the bacteria are provided with food in the form of sugars.

ROOT NODULE ON A PEA PLANT

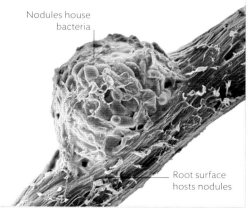

Nodules house bacteria

Root surface hosts nodules

Amanita muscaria

the fly agaric

Fungi play a vital role in the life of most plants. Many plants have evolved a symbiotic relationship with certain types of fungi, known as mycorrhizae, which include the fly agaric. This relationship gives plants access to extra water and nutrients in exchange for carbohydrates.

The fly agaric fungus is native to much of the northern hemisphere and is spreading throughout the southern hemisphere as well. All that most people ever see of this fungus is the colourful fruiting body, or mushroom. Its umbrella-like cap is bright red to orange-yellow and usually studded with white warts. After releasing its spores, the mushroom rots away.

Through its main underground body, the mycelium, the fly agaric plays a vital role in the ecology of coniferous and deciduous woodlands. This sprawling mass of hair-like structures (hyphae), spread throughout the soil, can form symbiotic (beneficial) relationships with the roots of a huge variety of trees, including pines, spruces, cedars, and birches. Whereas some fungi penetrate tree root cells, the fly agaric forms a sheath that covers the roots. The sheath not only protects the roots from infectious microorganisms but also helps to transfer both nutrients and water to the roots. In return, the tree supports the fungus by supplying it with sugars produced by photosynthesis. The best places to look for fly agarics, therefore, are near the base of the trees with which it associates.

Because this fungus can thrive with so many different trees, it is now appearing in places where it is not native, possibly by hitching a ride on the roots of tree seedlings that are destined for plantations. Some experts fear that it may compete with and drive out important local species of mycorrhizal fungi.

Toxic toadstool
The fly agaric brews an impressive cocktail of chemicals that both help it break down soil nutrients and protect it from being eaten.

The tiny white warts are remnants of a veil of tissue that protected the spore-bearing cap as it emerged from underground

The skirt-like ring of tissue around the stem, together with the cap, identify the fly agaric

Tip of the iceberg
A mushroom, like the one pictured here, is just a reproductive structure. The rest of the fungus lives underground and consists of countless hair-like structures called hyphae.

The hyphae envelop the roots of the host tree, and can grow between the root cells without penetrating the cell walls

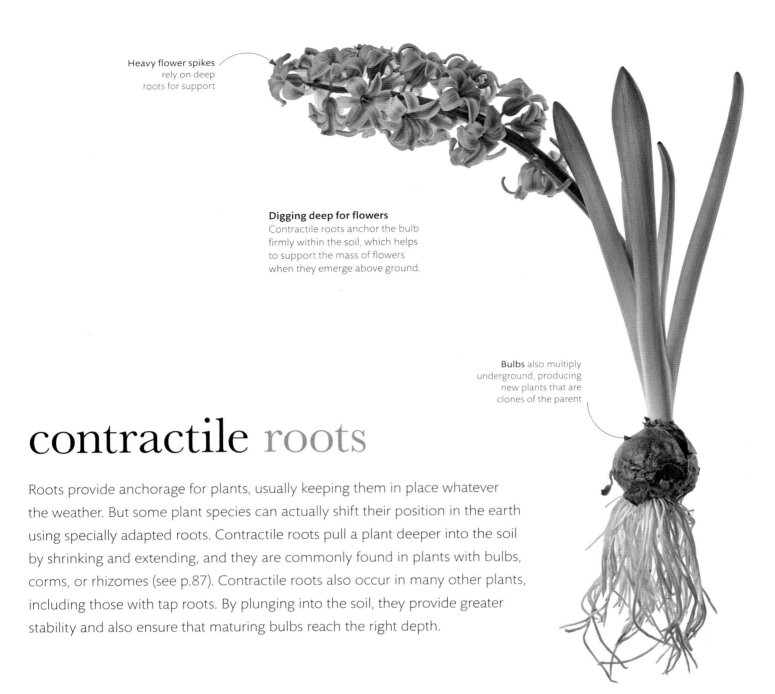

Heavy flower spikes
rely on deep
roots for support

Digging deep for flowers
Contractile roots anchor the bulb
firmly within the soil, which helps
to support the mass of flowers
when they emerge above ground.

Bulbs also multiply
underground, producing
new plants that are
clones of the parent

contractile roots

Roots provide anchorage for plants, usually keeping them in place whatever
the weather. But some plant species can actually shift their position in the earth
using specially adapted roots. Contractile roots pull a plant deeper into the soil
by shrinking and extending, and they are commonly found in plants with bulbs,
corms, or rhizomes (see p.87). Contractile roots also occur in many other plants,
including those with tap roots. By plunging into the soil, they provide greater
stability and also ensure that maturing bulbs reach the right depth.

FINDING THE RIGHT DEPTH
Bulbous plants begin life as seedlings
near the surface of the soil. However, if
the developing bulb stays near the soil
surface, then not only is it exposed to
freezing temperatures and the drying
rays of the sun, it is much more likely
to be eaten by an animal. For its
protection, contractile roots gradually
pull the developing bulb down into the
soil, where environmental conditions
are more stable. Contractile roots
widen before they extend, pushing the
surrounding soil aside to form a channel
through which the bulb is pulled.

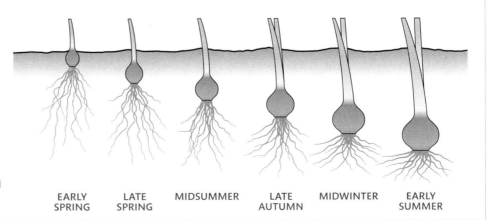

| EARLY SPRING | LATE SPRING | MIDSUMMER | LATE AUTUMN | MIDWINTER | EARLY SUMMER |

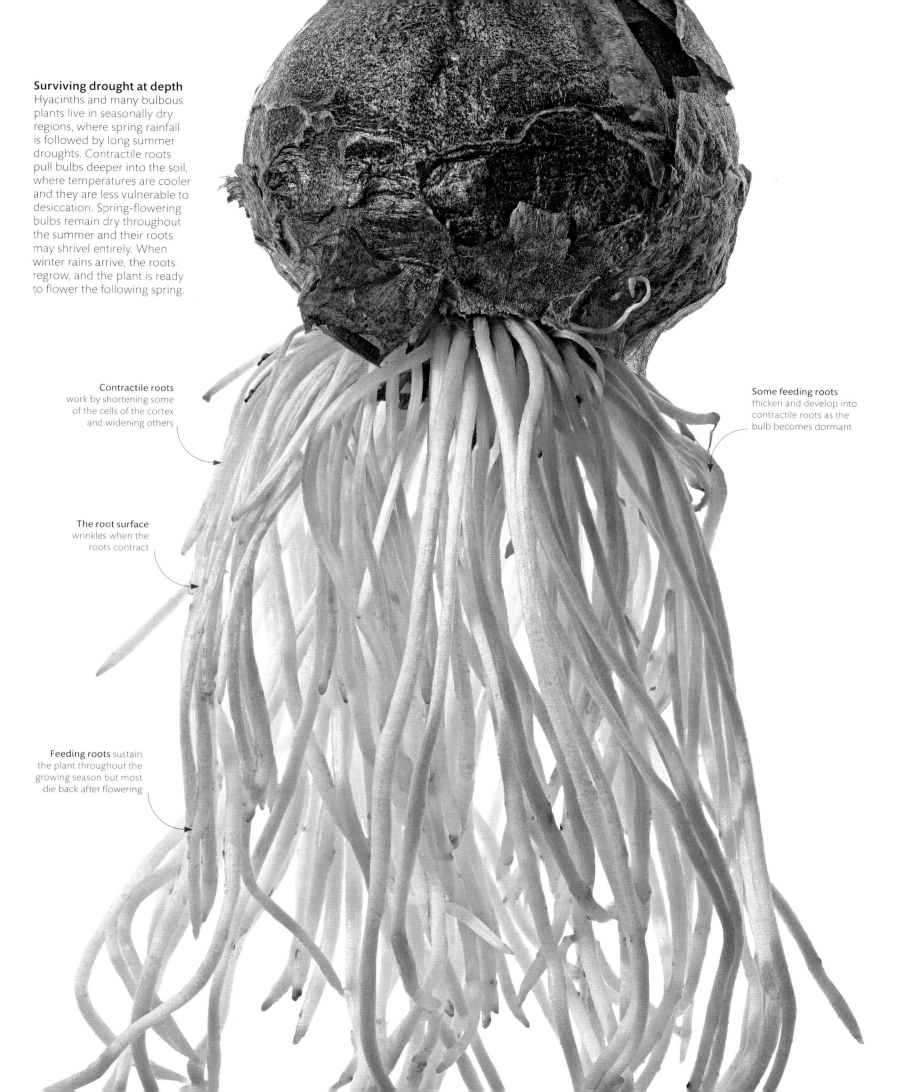

Surviving drought at depth
Hyacinths and many bulbous plants live in seasonally dry regions, where spring rainfall is followed by long summer droughts. Contractile roots pull bulbs deeper into the soil, where temperatures are cooler and they are less vulnerable to desiccation. Spring-flowering bulbs remain dry throughout the summer and their roots may shrivel entirely. When winter rains arrive, the roots regrow, and the plant is ready to flower the following spring.

Contractile roots work by shortening some of the cells of the cortex and widening others

The root surface wrinkles when the roots contract

Feeding roots sustain the plant throughout the growing season but most die back after flowering

Some feeding roots thicken and develop into contractile roots as the bulb becomes dormant

White berries attract birds, which distribute the seeds to other trees

aerial roots

Living high in the canopy, perched on the branches of trees, epiphytic plants need roots that hold them firmly in place. These aerial roots emerge along the stems and cling to any nearby surface. Unlike the roots of ground-dwelling terrestrial plants, which make direct contact with the soil, aerial roots can only absorb water from fog, mist, and rain. They are specially adapted to draw water from these sources. In some cases, aerial roots become green, and can manufacture food via photosynthesis (see p.129).

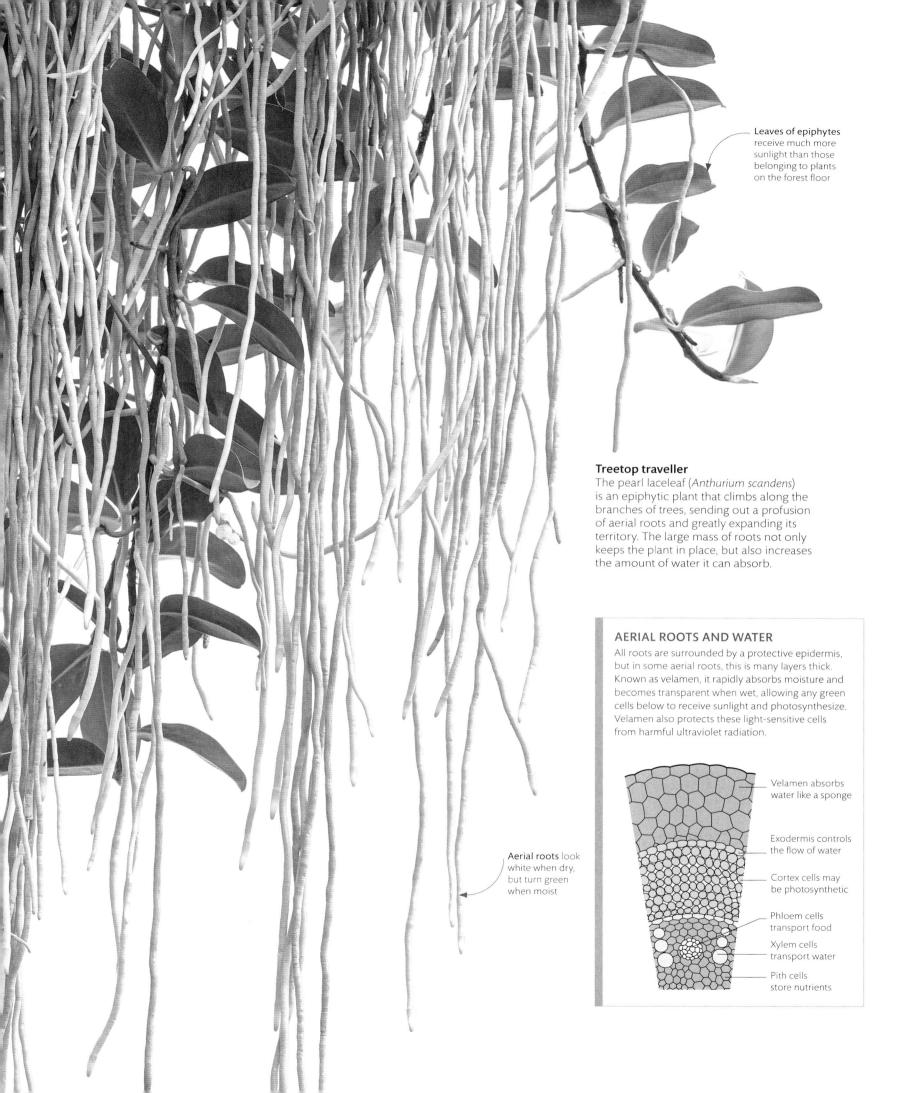

Leaves of epiphytes receive much more sunlight than those belonging to plants on the forest floor

Treetop traveller

The pearl laceleaf (*Anthurium scandens*) is an epiphytic plant that climbs along the branches of trees, sending out a profusion of aerial roots and greatly expanding its territory. The large mass of roots not only keeps the plant in place, but also increases the amount of water it can absorb.

AERIAL ROOTS AND WATER

All roots are surrounded by a protective epidermis, but in some aerial roots, this is many layers thick. Known as velamen, it rapidly absorbs moisture and becomes transparent when wet, allowing any green cells below to receive sunlight and photosynthesize. Velamen also protects these light-sensitive cells from harmful ultraviolet radiation.

Velamen absorbs water like a sponge

Exodermis controls the flow of water

Cortex cells may be photosynthetic

Phloem cells transport food

Xylem cells transport water

Pith cells store nutrients

Aerial roots look white when dry, but turn green when moist

Ficus sp.

the strangler fig

Some fig trees have evolved a way of life that involves growing on other trees and eventually throttling them. Many different *Ficus* species exhibit the strangler lifestyle. As unsavoury as this may sound, strangler figs are nonetheless vital components of tropical forests.

A strangler fig starts life as a tiny seed deposited on a tree branch by an animal. After germinating, the seedling's roots dig into whatever deposits have accumulated on the branch. Over time, those roots go in search of more nutrients, snaking their way down the trunk of their host. As soon as they touch the ground, the fig switches from being a harmless epiphyte into a lethal lodger. Like woody webbing, the roots of the fig grow larger and larger until they kill their host by strangulation.

Initially, strangler figs may give their host trees some protection via their roots, preventing them from being uprooted in tropical storms – but this benefit only lasts until the strangler throttles its host.

Hollow victory
When the dead tree host rots away, all that remains is the strangler fig, whose roots form a cast of the tree that once supported it. The hollow space inside is a safe habitat for birds, insects, and bats.

Strangler figs produce copious amounts of fruits. Each fruit is a fleshy syconium, basically an inverted inflorescence with its flowers on the inside. Tiny fig wasps are responsible for pollination. A female enters the syconium through a small hole and lays her eggs near the fig ovules. As she does so, she spreads pollen from another fig among the female flowers. The wasps are born, feed, and mate inside the fig. Impregnated females then collect pollen from male flowers and, leaving the male wasps behind, fly off to another fig, where the process starts once again. Enough flowers are fertilized for the fruit to contain plenty of viable seeds. With luck, a tree-living animal will eat the fruit and deposit its seeds on the branch of another host tree, continuing the cycle.

This lifestyle enables the fig seeds to germinate close to the treetop canopy, which receives far more sunlight than the forest floor, where it would struggle to survive.

Sweet fruits
The energy-rich fruits of strangler figs are relished by a wide range of creatures. They carry the seeds far from the parent tree, passing them out in their droppings.

Fig fruits are packed with seeds, which remain viable after the fruits have been eaten and digested

Air plant leaves are covered with silver, shield-like scales that absorb water from the hot, steamy air of the forests where they grow

living on air

Air plants are members of the pineapple family (Bromeliaceae), and they are named after their apparent ability to thrive on fresh air alone. Unlike most plants, which rely on roots to take up water from the soil, they can absorb moisture from the air through scales on their leaves. Air plants belong to the genus *Tillandsia*, and although most species do have roots, these serve largely to fix them into position on a tree branch or rock.

Life in the canopy
Epiphytes live perched on other plants. They are not parasites and do not feed from their hosts, but they benefit from living on branches high up in the rainforest, where they receive much more sunshine than they would on the forest floor. Tillandsias are not the only epiphytic plants – many ferns, orchids, and other bromeliads also live high in the treetops.

Colourful bracts enclose *Tillandsia* flowers and attract hummingbirds and other pollinators

The leaf scales become transparent when wet, turning the silvery leaves green

Wiry roots

Silky hairs
Air plant flowers produce numerous seeds equipped with fine, silky hairs. These enable the seeds to be blown away on the wind to new branches where they can lodge and grow.

TILLANDSIA TENUIFOLIA

AIR PLANT HOSTS
Tillandsia are usually found growing on branches or rocks. Smaller species can adhere to the flimsiest of twigs and in urban areas they will even fasten themselves to telegraph poles and overhead cables. It is easiest for air plant seeds to wedge themselves into the crevices of rough surfaces, such as tree bark.

TILLANDSIA TECTORUM

TILLANDSIA IONANTHA

parasitic plants

While most plants produce their own food via photosynthesis, there are some that cheat. Parasitic plants prey upon other plant species, stealing water and carbohydrates from them by penetrating their host's tissues using modified roots known as haustoria. Some plants, such as mistletoe, attach themselves to stems and branches, while others live on the root systems of their hosts. Some parasites can only live when they are connected to a host, but others are able to survive independently.

Red bartsia
(*Odontites vulgaris*)
parasitizes a range
of different species

Hemiparasites
Although parasites like yellow rattle and red bartsia have green leaves and can make their own food, they steal water from their hosts. These "hemiparasites" may also pilfer carbohydrates, to supplement their own supply.

Yellow rattle
(*Rhinanthus minor*)
is a hemiparasite and
can survive without
a host plant

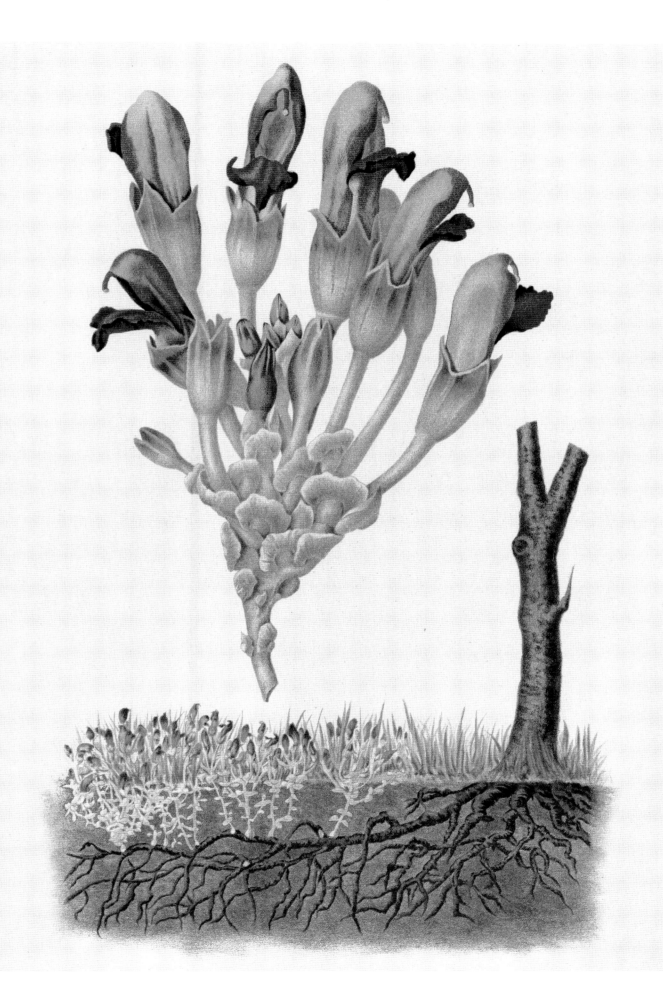

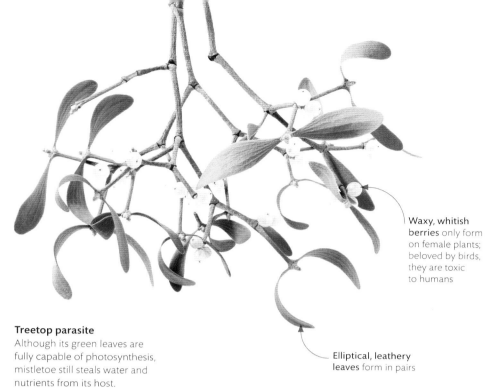

Waxy, whitish berries only form on female plants; beloved by birds, they are toxic to humans

Treetop parasite
Although its green leaves are fully capable of photosynthesis, mistletoe still steals water and nutrients from its host.

Elliptical, leathery leaves form in pairs

Viscum album

mistletoe

Engrained in myth and folklore, and an excuse for a kiss at Yuletide, mistletoe has a fascinating biology. This parasitic plant lives on various deciduous tree species. It may stunt growth and cause deformities, but it rarely kills its host; in fact, if the host tree dies, so does the mistletoe.

Instead of growing its own roots, mistletoe develops specialized structures called haustoria that penetrate the vascular tissues of its host tree to absorb water and nutrients. Mistletoe is slow-growing, so healthy host trees can tolerate a few mistletoe plants without serious ill effects. Trees with heavy infestations, however, may be weakened and are less likely to survive additional stresses such as disease, drought, or temperature extremes.

Birds are crucial to the spread of mistletoe. After mistletoe's small flowers are pollinated, many white-to-yellow berries are produced. Birds love to feed on the berries, but they can only digest the soft pulp, so they either excrete or squeeze out the poisonous seed while eating the fruit. The seeds then stick to the bird's face, and are wiped off on to a branch. There, the sticky coating hardens, and glues the seeds to the branch. The seeds then send out their haustoria into the host, to complete the plant's life cycle.

Although parasitic, mistletoes play an important role wherever they grow. There are many species of mistletoe, and each one is a vital food source for birds and insects. Those animals, in turn, attract more wildlife, and it is now clear that mistletoes help to increase biodiversity in their native habitats. Moreover, the fact that they prefer certain trees to others helps to prevent the species that they colonize from becoming dominant to the detriment of other types of tree.

Mistletoe in winter
Evergreen mistletoe is easiest to spot during winter, when dense clusters up to 1 m (3 ft) wide can be seen dangling from naked trees. Each cluster is an individual mistletoe plant made up of many regularly forking branches.

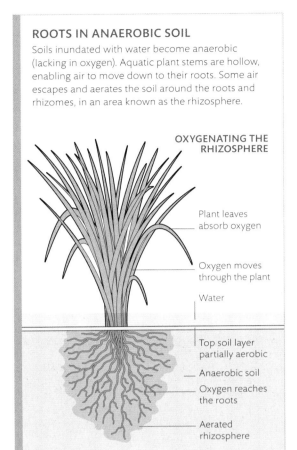

ROOTS IN ANAEROBIC SOIL

Soils inundated with water become anaerobic (lacking in oxygen). Aquatic plant stems are hollow, enabling air to move down to their roots. Some air escapes and aerates the soil around the roots and rhizomes, in an area known as the rhizosphere.

OXYGENATING THE RHIZOSPHERE

Plant leaves absorb oxygen

Oxygen moves through the plant

Water

Top soil layer partially aerobic

Anaerobic soil

Oxygen reaches the roots

Aerated rhizosphere

Prehistoric plant
Commonly known as horsetails, *Equisetum* species have grown in damp areas, including those that are seasonally flooded, for over 300 million years. Today's varieties are miniature versions of their ancestors, which could grow up to 1 m (3 ft) in diameter.

roots in water

All plant cells require a source of oxygen if they are to survive, so for aquatic plants, getting air down to their submerged roots is crucial. Hollow channels with a spongelike structure, called aerenchyma, run through the leaf, stem, and root tissues of aquatic plants. They allow air to flow through the cavities from above the water's surface down to the roots.

Horsetail stems are rich in silica, a mineral that makes them rough and unappetizing, deterring herbivores

Firm, vertical ridges give support to the hollow stems

Tiny leaves shrink and fuse to form toothed, collar-like sheaths at the nodes

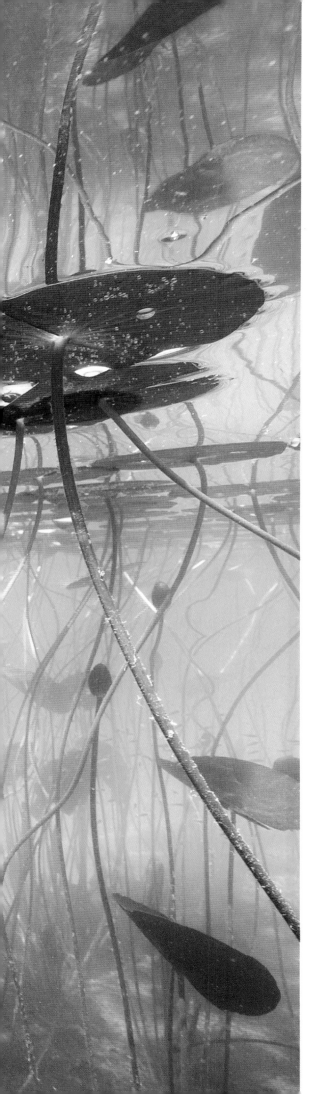

Nymphaea sp.

water lilies

The leaves of water lilies (*Nymphaea*) lie upon the water's surface with the utmost grace and their flowers punctuate the scene with subtle colours. DNA studies show that water lilies have one of the oldest lineages of any group of flowering plants.

There are over 60 known species of *Nymphaea*, and they can be found growing in both tropical and temperate climates. Related to all other angiosperms except for the obscure *Amborella*, water lilies are an evolutionary success story, millions of years in the making. Look out over a lily pond and it is easy to mistake water lilies for floating plants. In fact, the leaves of *Nymphaea* plants are actually attached to long, slender stems that arise from a thick rhizome buried deep in the mud, and they are able to float with the help of large air sacs packed between their cells.

Aquatic elegance
Water lilies have become prized additions to water gardens the world over. Where they escape into the wild, they often crowd out the native plants, upsetting the delicate balance of many aquatic ecosystems.

When a water lily comes into bloom, the female parts of the flower (the stigma) mature first. The bowl-shaped stigma is filled with a thick, sticky fluid containing compounds that attract insects such as bees and beetles. As they crawl into the centre in search of their sweet reward, pollen from other water lily blooms is washed off the insect and into the stigma, pollinating the flower. Some insects perish in the process and drown: it makes no difference to the water lily whether or not its pollinator survives.

After the first day, the flower stops producing fluid, and the stamens become active for a day or two, releasing pollen for insects to collect and deliver to other blooms, continuing the cycle. When the flower closes for the last time, its stem recoils, drawing the bud back underwater. This positions the developing seeds closer to the mud, where they can germinate.

The male and female parts of the flower mature at different times, reducing the chances of self-pollination

Opening bud
Water lily flowers only emerge above the water to be pollinated. Once this has happened, they retreat back below the surface, to give their seeds a better chance to germinate.

HOW ROOTS BREATHE

Tidal mangrove roots are submerged and exposed twice daily, and the soil they occupy is anaerobic, containing little or no oxygen. In order to breathe, some species develop upright extensions of the root system called pneumatophores, which act like snorkels. They absorb air through pores, known as lenticels, on their surface and transport it to the roots.

Lenticels absorb air

Sea water

Sediment

PNEUMATOPHORES

breathing roots

Shallow coastal waters represent one of the most challenging habitats for plants. Tidal flows and storm surges uproot them, especially as the soft sediment of the sea floor provides only minimal anchorage. Salty water also dries out plant tissues, and roots and stems are starved of oxygen because they are submerged. Mangroves are among the few groups of trees and shrubs to thrive in such environments, and the forests they form protect coastal communities from storms and erosion.

Pneumatophores are root extensions that push their way through the swampy ground

Roots for anchorage

In order to resist the pull of the tide and to survive tropical storms, some mangroves develop extensive networks of prop roots. These not only fix the plant in the shallow soil, they also slow the flow of water, with more sediment retained around the roots.

Breathing air, excluding salt

Mangrove roots can breathe air during low tide, through their pneumatophores. Some species can also exclude salt; the root membranes work like a filter, allowing water to enter while keeping out the harmful salt.

mangroves

Living in saline water is an immense challenge for plants – but one that mangroves have successfully overcome. Of all the trees and shrubs referred to as mangroves, relatively few are "true mangroves" – meaning those that live exclusively in salt water habitats, such as the various *Rhizophora* species.

The dehydrating effects of salt and the difficulties of obtaining fresh water make coastal conditions inhospitable. *Rhizophora* mangroves get around this problem by filtering out the salt. The trees sit on top of long, spindly prop roots that give them their characteristic appearance, and these prop roots are the key to the mangroves' success. Water entering the roots passes through a series of cellular filters that remove salts, giving the trees access to an endless supply of fresh water. By always keeping parts of their roots above water, the mangroves are also able to exchange carbon dioxide and oxygen gas, and thus avoid suffocation (see p.53).

Rhizophora mangroves play a vital role in sustaining coastal communities – both human and animal. Their roots hold on to sand, slow erosion, and reduce the intensity of waves, protecting and building up shorelines that would otherwise be washed away. Forests of mangroves also help to shield settlements and wildlife from hurricanes and tropical storms, and provide important feeding and nesting sites for many types of bird.

Rhizophora mangroves rely on the ebb and flow of the sea to colonize new sites. These trees are viviparous, which means that their seeds germinate on the branch before dispersal. The torpedo-shaped seedlings either plunge into the sand at the base of the parent tree or float away on the tide and, with luck, wash up on a faraway beach ready to start a new forest.

Mangrove forest at high tide
Countless fish species spawn among the tangled mass of *Rhizophora* roots, and their offspring grow up in the shelter of these forests.

The tops of **roots** always remain above the water, even at high tide

Low-tide mangrove
The prop roots of this red mangrove (*Rhizophora mangle*) curve down into the sand at a salt-water lagoon in the Bahamas. The uppermost part of the roots are never submerged; thus gas exchange can take place continuously, enabling photosynthesis and respiration.

stems and branches

stem. the main body or stalk of a plant or shrub, usually rising above ground but occasionally subterranean. **branch.** a limb that grows out from the trunk of a tree, or from the main stem of a plant.

types of stem

Stems are the skeletons of plants, supporting and connecting roots, leaves, flowers, and fruits. They conceal a circulatory system that moves water and food around the whole plant. Stem structures show huge variety, from majestic trees and arching vines to spreading carpets and underground rhizomes, and they range in size from the wiry stems of miniature mosses to the massive trunks of forest redwoods.

Hard and soft stems

Secondary thickening is the process by which stems produce woody tissue, allowing them to become larger and stronger. Many plants, however, never become woody; their soft, herbaceous stems only last a single growing season.

Internodes contain a spongy pith and a sugar-rich liquid

Strongly upright sugar cane stems can reach up to 5 m (15 ft) tall

Tough stems are essential for this tall grass to support itself

Young ivy stems are soft and flexible to aid climbing; they become woody with age

Stems bearing flowers or flower heads (or later fruits) are called peduncles

ICELANDIC POPPY
(*PAPAVER NUDICAULE*)

SUGAR CANE
(*SACCHARUM OFFICINARUM* 'KO-HAPAI')

COMMON IVY
(*HEDERA HELIX*)

Bark protects woody stems from damage, water loss, and destructive insects

From stems to branches
As the layers of xylem (woody tissue) build up, the stems become much stronger and thicker. The tallest plants in the world are trees with woody stems.

Woody stems are durable enough to survive for many years

The contorted stems of the corkscrew hazel are a genetic abnormality, first discovered growing wild in a hedgerow

CORKSCREW HAZEL
(*CORYLUS AVELLANA* 'CONTORTA')

SILVER BIRCH
(*BETULA PENDULA*)

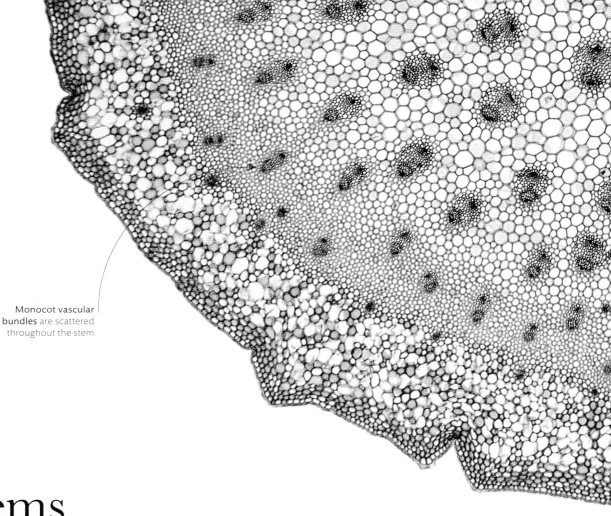

Monocot vascular **bundles** are scattered throughout the stem

inside stems

All stems have two key functions: support and transport. They hold the leaves aloft, allowing them to absorb sunlight, and then transport the carbohydrates that are manufactured there around the plant. Water and minerals are also carried up from the roots via the woody xylem tissue, made of dead cells, while food and other materials are transported around the stems through the living phloem cells.

Stems and vascular bundles
Inside the stem, xylem and phloem cells are packed together into vascular bundles. Among flowering plants, these are arranged differently within the stem depending on whether the plant is a monocot or a eudicot (see p.15). In monocots, bundles are scattered throughout the core of the stem, but in other flowering plants, and all eudicots, the bundles are arranged in a circle. This is clearly visible in trees: over time, most trees develop rings in their trunk, but monocot trees, such as palms, never do.

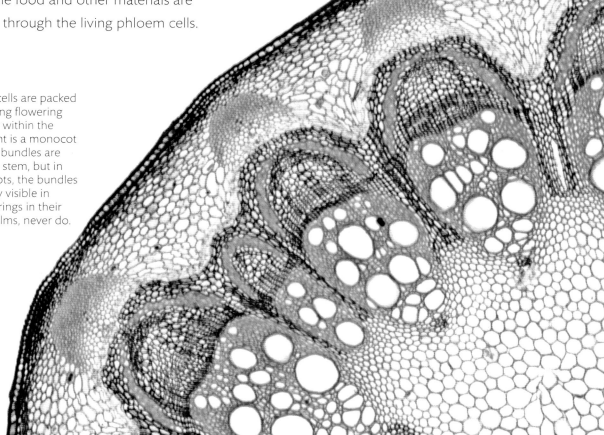

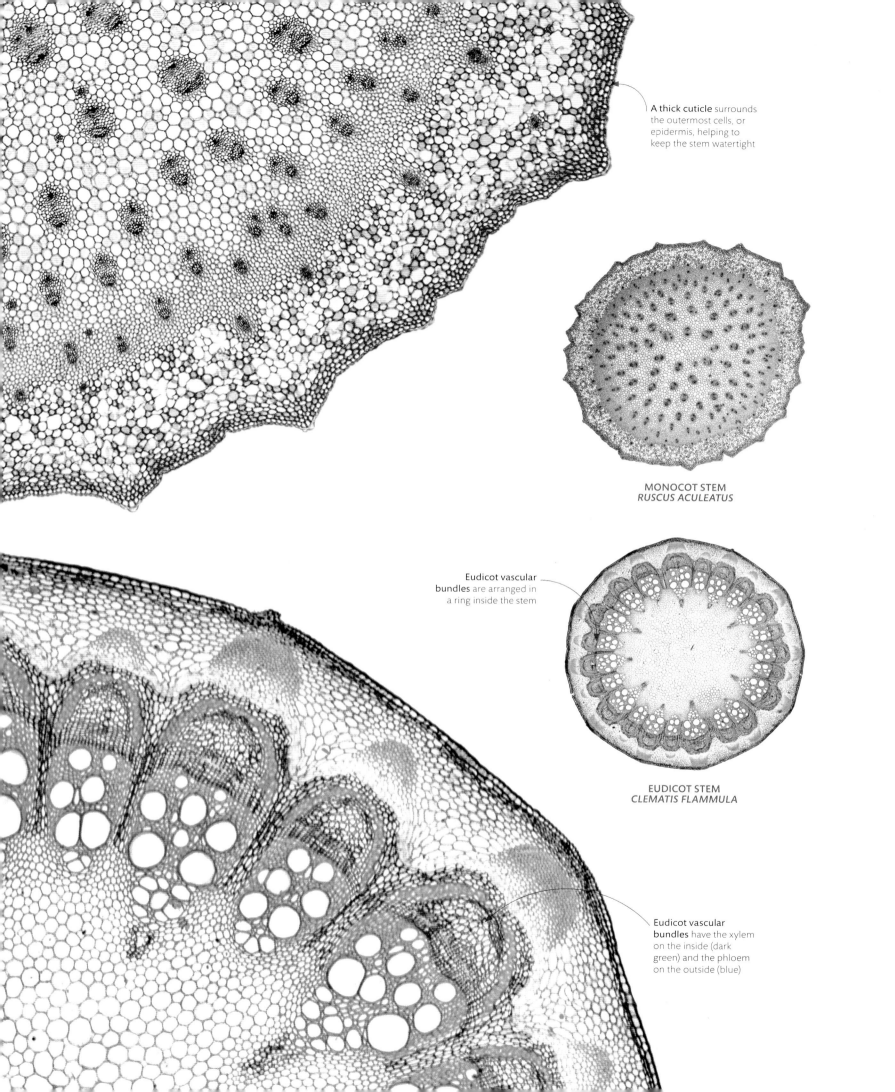

A thick cuticle surrounds the outermost cells, or epidermis, helping to keep the stem watertight

MONOCOT STEM
RUSCUS ACULEATUS

Eudicot vascular bundles are arranged in a ring inside the stem

EUDICOT STEM
CLEMATIS FLAMMULA

Eudicot vascular bundles have the xylem on the inside (dark green) and the phloem on the outside (blue)

Great Piece of Turf, 1503

Albrecht Dürer's meticulous watercolour study of an everyday clump of grasses and weeds gains its power from perspective – the viewer is down at ground level with the insects and small creatures that inhabit the turf. Artfully naturalistic against a plain background, the painting features a range of perfectly rendered plant species, including cock's-foot, creeping bent, smooth meadow-grass, daisy, dandelion, germander speedwell, greater plantain, hound's-tongue, fool's watercress, and yarrow.

natural renaissance

Careful studies
Leonardo da Vinci's exquisite chalk and crayon sketches of plants and trees were often created as preparatory studies for his larger works, but they were also an important part of his investigations into botanical science.

Throughout the 200 years of the Renaissance, the scope of intellectual curiosity and human creativity seemed boundless. Major artists studied anatomy for sculpture and figure work, mathematics to solve problems of linear perspective, and the natural world in order to reproduce plant life and landscapes with complete accuracy. Their botanical sketches and watercolours are celebrated today for their naturalism.

Historically, plants were studied and illustrated in herbals (see pp.140–41) for identification. In the Middle Ages, artists depicted flowers, such as the lily (purity), to add symbolic meanings to religious paintings. This can be seen in the early works of Italian artist Leonardo da Vinci. However, in the late 15th century, the rediscovery of nature had a huge influence on Renaissance art. It inspired Leonardo to underpin his great paintings with close studies of plant species and scientific investigations into botanical processes, introducing a new, naturalistic look.

His work was an inspiration to another Renaissance giant, the German artist Albrecht Dürer, who was a master of oils, woodcuts, and engraving.

Dürer was renowned for his messianic self-portraits and visionary works on mythical and religious themes, but his private work was entirely different. The handful of quiet, perfectly observed watercolour studies of nature, which he probably intended to use to add realism to his religious paintings, include this uncultivated slice of summer meadow, a teeming microcosm of the natural world.

" … I realized that it was much better to insist on the genuine forms of nature, for simplicity is the greatest adornment of art. "

ALBRECHT DÜRER, *LETTER TO REFORMATION LEADER PHILIPP MELANCHTHON*

The bark splits as the trunk expands, but new layers form underneath

Phloem tissue lies just beneath the bark and transports food around the tree

Vascular cambium produces new xylem layers

Each pale ring is composed of early wood, which forms when the tree starts to grow in spring

Dark rings are late wood, which forms later in the year, just before the tree becomes dormant

tree trunks

A cross section of a tree trunk provides an unparalleled opportunity to peer into the past. Each year, the trunk – a woody stem – develops a new layer of tissue, the thickness of which is determined by environmental conditions. Good conditions lead to vigorous growth, which creates a wide ring. Stress caused by extreme temperatures or drought results in thin rings. Studying these rings provides a glimpse into the weather conditions of the past.

Strong support
The column-like shape of most tree trunks provides a physical support for the tree's framework of branches and thousands of leaves. Tree trunks can grow massively tall and are incredibly strong. They can hold themselves upright without needing a structure to climb around or wrap themselves over.

STRUCTURE OF A TREE TRUNK

Within the woody stems of trees and some flowering plants are rings of xylem and phloem, the tissues that transport water and food around the plant. A thin layer of phloem sits just under the bark, while multiple layers of xylem form the growth rings visible when a tree is felled. Every year, tissue known as vascular cambium produces new layers of xylem on top of the previous year's layer; the layers can be counted to estimate the age of the trunk. The younger, outer layers of xylem continue to transport water and are known as sapwood; the older layers on the inside gradually become blocked and form heartwood. On the outside of the woody layers, cork cambium produces new bark, to cover and protect the expanding tree trunk.

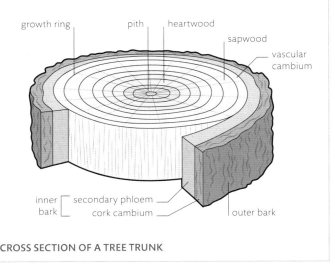

growth ring   pith   heartwood

sapwood

vascular cambium

inner bark ⎱ secondary phloem
bark ⎰ cork cambium

outer bark

CROSS SECTION OF A TREE TRUNK

CORKY
Quercus suber

STRIATE
Acer pensylvanicum

RIDGED
Castanea sativa

LENTICELLATE
Prunus serrula

SCALY
Pinus sp.

SPINY
Ceiba speciosa

FLAKING
Platanus sp.

PEELING
Eucalyptus gunnii

PEELING IN STRIPS
Carya ovata

SMOOTH
Betula populifolia

FISSURED
Liriodendron tulipifera

PAPERY
Acer griseum

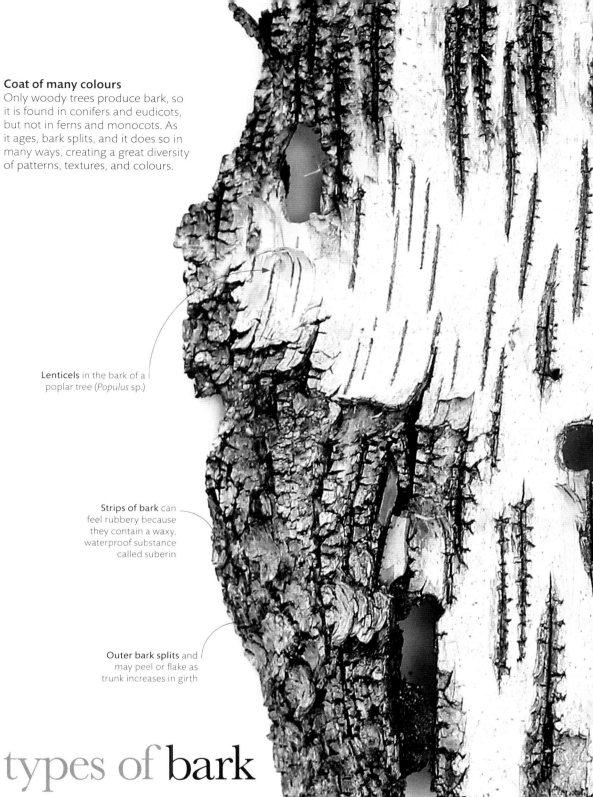

Coat of many colours
Only woody trees produce bark, so it is found in conifers and eudicots, but not in ferns and monocots. As it ages, bark splits, and it does so in many ways, creating a great diversity of patterns, textures, and colours.

Lenticels in the bark of a poplar tree (*Populus* sp.)

Strips of bark can feel rubbery because they contain a waxy, waterproof substance called suberin

Outer bark splits and may peel or flake as trunk increases in girth

types of bark

The protective "skin" of woody plants, bark keeps out invading insects, bacteria, and fungi, and retains precious moisture. It can also defend a tree from fire, and trees that shed layers of bark deter clinging vines and epiphytes from taking hold. Beneath the bark there are two important layers of dividing cells called cambium. They are relatively shallow, and damage to them can impede growth and, in extreme cases, kill the tree.

Populus tremuloides

the quaking aspen

With striking white trunks and leaves that tremble in the wind, few sights are as captivating as a grove of quaking aspen trees in autumn. This remarkable species has the widest distribution of any tree in North America, ranging from Canada all the way south into Mexico.

The quaking aspen lives for an extremely long time, although not in the traditional sense. Each tree is either male or female, making sexual reproduction possible, but it rarely reproduces by seed. Instead, once a tree is established, it grows multiple offshoots from its roots. Each offshoot is capable of growing into a new tree, so entire aspen groves can be made up of clones of a single tree. Over time, the trees eventually die, but the root stock itself can continue to produce new trees for hundreds, or even thousands, of years. The largest known grove of aspen clones, the Pando grove in Utah, is 80,000 years old and covers 40 hectares (100 acres).

The stark white bark helps to protect the tree from overheating and reduces the risk of sunscald during winter, when the bark thaws and then freezes again.

By reflecting most solar rays, the tree is able to maintain lower temperatures on sunny winter days. A close look at the bark reveals a greenish tinge: this is photosynthetic tissue, so even before its leaves emerge in spring, the tree is busy harvesting sunlight for photosynthesis.

The tree's habit of cloning itself enables stands to regrow even after forest fires. In fact, fire is essential for maintaining the quaking aspen's habitat. Without cleared land, aspens would eventually be shaded out by trees such as conifers.

Forest growth
The uniformity seen in this stand of quaking aspens suggests it might be a clonal group. Aspens respond rapidly when the ground is cleared; as more light reaches the soil, they quickly send up new, fast-growing shoots.

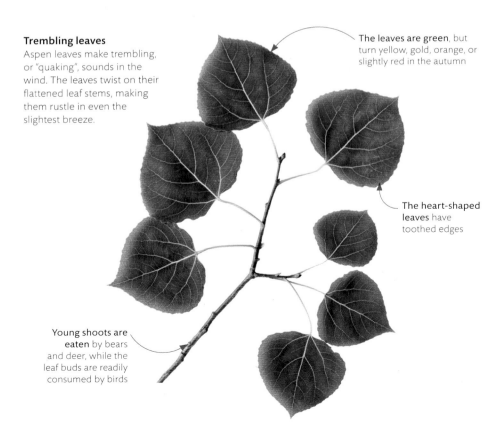

Trembling leaves
Aspen leaves make trembling, or "quaking", sounds in the wind. The leaves twist on their flattened leaf stems, making them rustle in even the slightest breeze.

The leaves are green, but turn yellow, gold, orange, or slightly red in the autumn

The heart-shaped leaves have toothed edges

Young shoots are eaten by bears and deer, while the leaf buds are readily consumed by birds

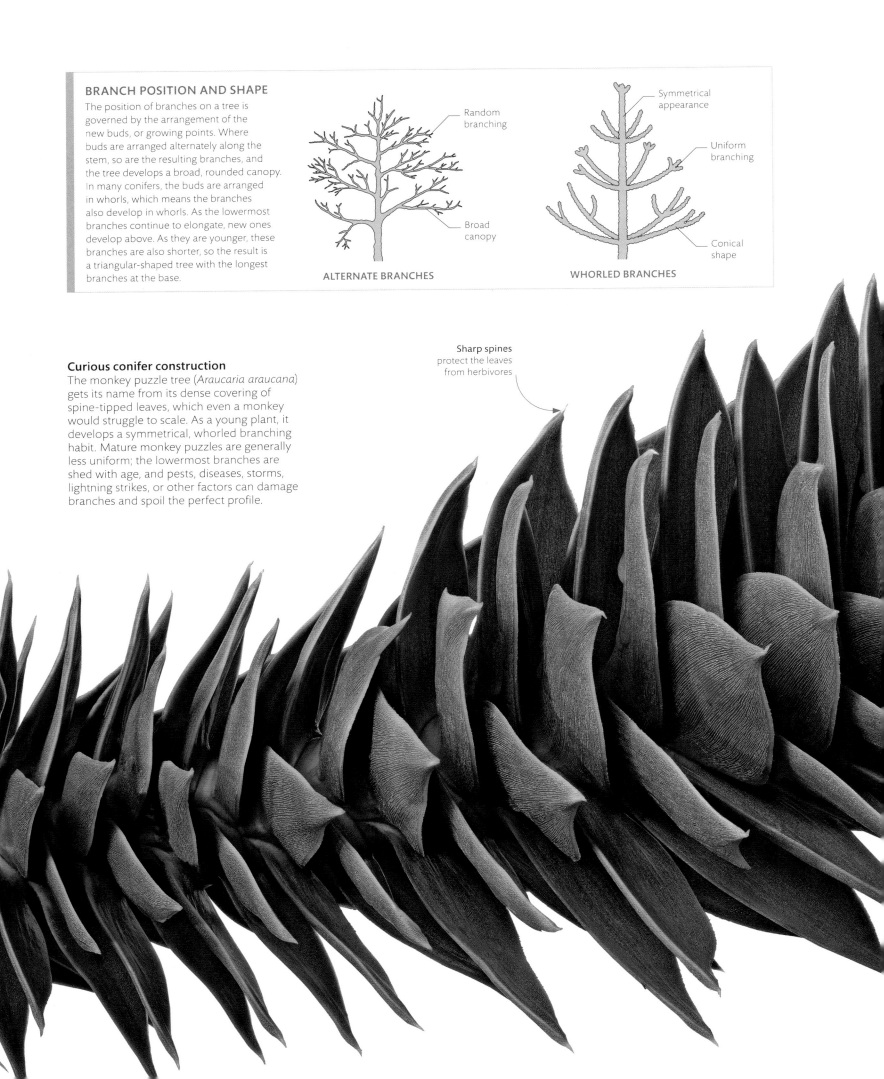

BRANCH POSITION AND SHAPE

The position of branches on a tree is governed by the arrangement of the new buds, or growing points. Where buds are arranged alternately along the stem, so are the resulting branches, and the tree develops a broad, rounded canopy. In many conifers, the buds are arranged in whorls, which means the branches also develop in whorls. As the lowermost branches continue to elongate, new ones develop above. As they are younger, these branches are also shorter, so the result is a triangular-shaped tree with the longest branches at the base.

Random branching

Broad canopy

ALTERNATE BRANCHES

Symmetrical appearance

Uniform branching

Conical shape

WHORLED BRANCHES

Curious conifer construction

The monkey puzzle tree (*Araucaria araucana*) gets its name from its dense covering of spine-tipped leaves, which even a monkey would struggle to scale. As a young plant, it develops a symmetrical, whorled branching habit. Mature monkey puzzles are generally less uniform; the lowermost branches are shed with age, and pests, diseases, storms, lightning strikes, or other factors can damage branches and spoil the perfect profile.

Sharp spines protect the leaves from herbivores

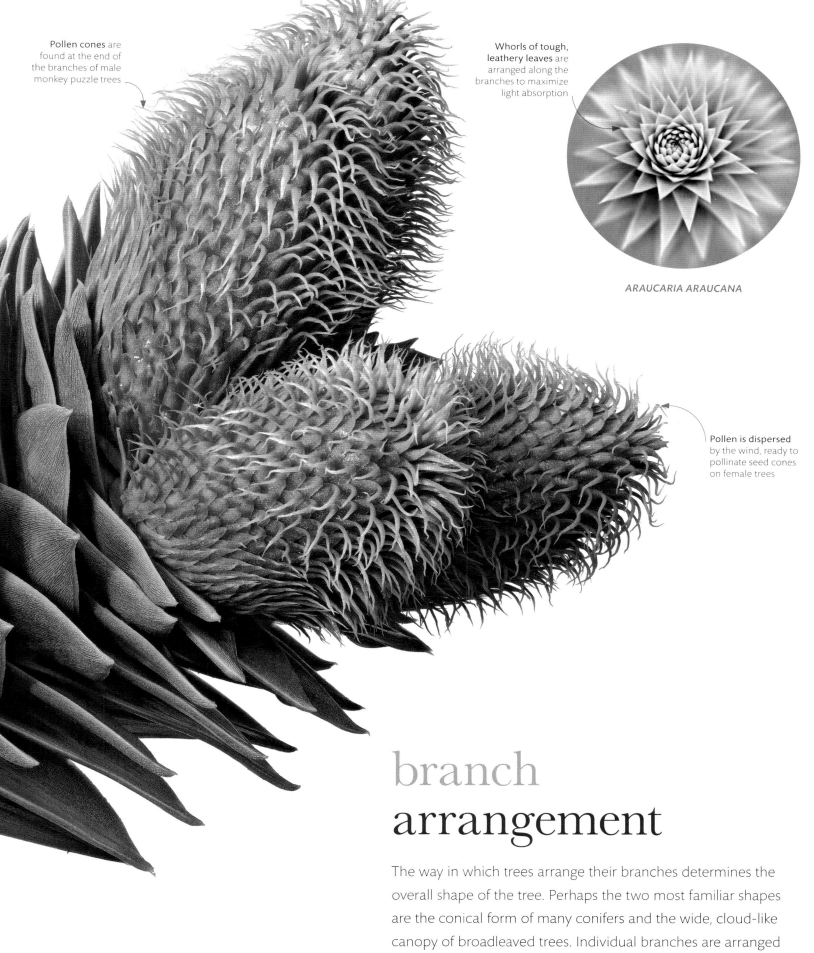

Pollen cones are found at the end of the branches of male monkey puzzle trees

Whorls of tough, leathery leaves are arranged along the branches to maximize light absorption

ARAUCARIA ARAUCANA

Pollen is dispersed by the wind, ready to pollinate seed cones on female trees

branch
arrangement

The way in which trees arrange their branches determines the overall shape of the tree. Perhaps the two most familiar shapes are the conical form of many conifers and the wide, cloud-like canopy of broadleaved trees. Individual branches are arranged to maximize the amount of light that the leaves can receive.

Terminal buds form at the top of the twig or stem

Scaled buds may be covered by resin for extra protection in winter

Pseudo-terminal buds form when branches die back to an axillary bud

Flower buds tend to be larger than leaf buds

Axillary (or lateral) buds form where the last year's leaf joined the stem

Leaf scars are sometimes left by the old leaves when they fall

Unscaled buds have no scales, but may still be protected by hairy bracts

Juglans mandshurica var. sieboldiana

Aesculus hippocastanum

Quercus frainetto

Magnolia campbellii

TERMINAL BUD

AXILLARY BUD

PSEUDO-TERMINAL BUD

UNSCALED BUD

winter buds

Leaf buds vary enormously and are highly distinctive, both in shape and in the way in which they grow. Along with a tree's shape, they are a vital aid to identifying a tree during winter. Examining how buds grow along a stem is revealing. They can form opposite each other or on alternate sides of a stem, arising around it at intervals. Bud scales, which protect the developing leaves and flowers, also vary in shape, colour, and number.

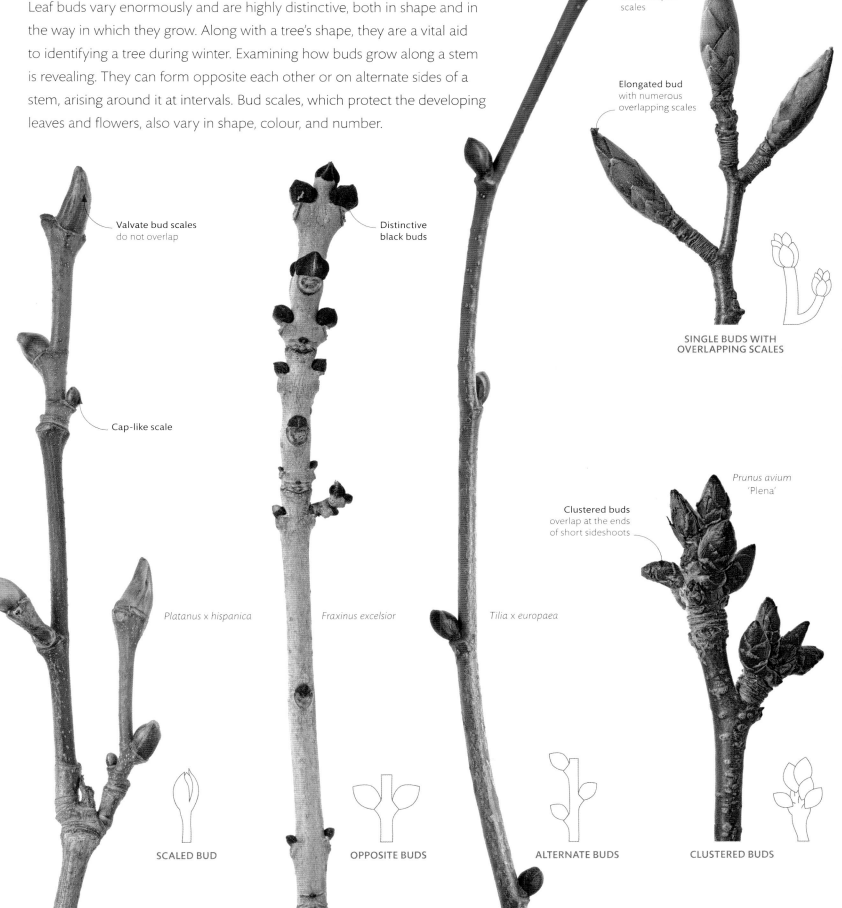

Valvate bud scales
do not overlap

Cap-like scale

Platanus x *hispanica*

SCALED BUD

Distinctive
black buds

Fraxinus excelsior

OPPOSITE BUDS

Smooth bud
with unequal
scales

Elongated bud
with numerous
overlapping scales

Fagus sylvatica

**SINGLE BUDS WITH
OVERLAPPING SCALES**

Tilia x *europaea*

ALTERNATE BUDS

Clustered buds
overlap at the ends
of short sideshoots

Prunus avium
'Plena'

CLUSTERED BUDS

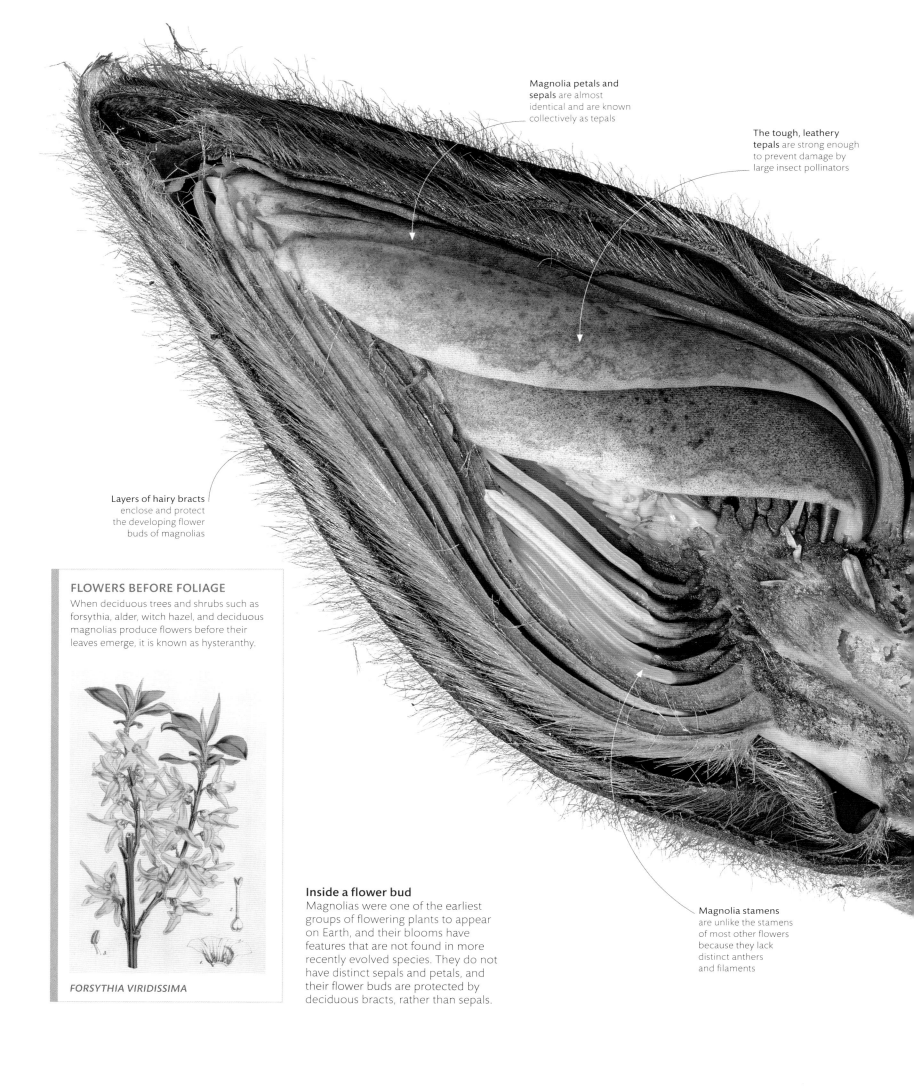

Magnolia petals and sepals are almost identical and are known collectively as tepals

The tough, leathery tepals are strong enough to prevent damage by large insect pollinators

Layers of hairy bracts enclose and protect the developing flower buds of magnolias

FLOWERS BEFORE FOLIAGE

When deciduous trees and shrubs such as forsythia, alder, witch hazel, and deciduous magnolias produce flowers before their leaves emerge, it is known as hysteranthy.

FORSYTHIA VIRIDISSIMA

Inside a flower bud

Magnolias were one of the earliest groups of flowering plants to appear on Earth, and their blooms have features that are not found in more recently evolved species. They do not have distinct sepals and petals, and their flower buds are protected by deciduous bracts, rather than sepals.

Magnolia stamens are unlike the stamens of most other flowers because they lack distinct anthers and filaments

Scales and scars
Magnolia stems are distinctive, even when bare of leaves. The hairy bracts (sometimes referred to as bud scales) that protect the flower buds, and the circular scars beneath them, are easily recognized. The shield-shaped leaf scars are another characteristic.

Bracts either fall as the flower opens, or they are shed prior to flowering

Crusty lichens grow on the long-lived stems and branches

Silky bract hairs may be silver or fawn, or sometimes absent

Distinctive scars are left on the stems by fallen leaves

insulated buds

As well as leaf buds, the woody stems of trees and shrubs also bear the buds of next year's flowers. Deciduous magnolias form their flower buds in late summer and autumn, remaining dormant in the winter. It is usually possible to distinguish between flower and leaf buds by their shape or size.

Buds that develop into foliage are much smaller than the flower buds

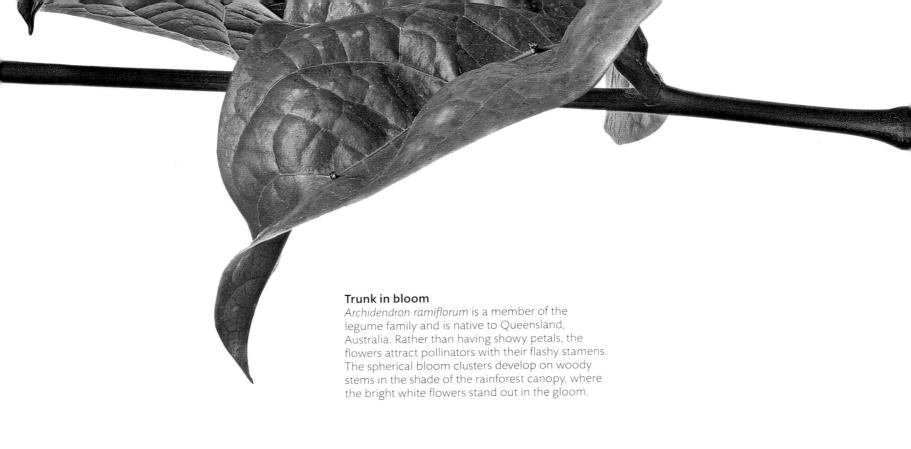

Trunk in bloom
Archidendron ramiflorum is a member of the legume family and is native to Queensland, Australia. Rather than having showy petals, the flowers attract pollinators with their flashy stamens. The spherical bloom clusters develop on woody stems in the shade of the rainforest canopy, where the bright white flowers stand out in the gloom.

cauliflorous plants

Flowers and fruit usually develop on new shoots, but in some trees and shrubs, the blooms erupt directly from the woody trunk and primary branches. Known as cauliflory, this strategy is more common in the tropics than in cooler regions. The reasons why some plants adopt cauliflory remain mysterious, but it may be an evolutionary adaptation to allow animals that live lower down in the forest canopy easy access to the flowers and fruits. Despite the name, cauliflowers are not cauliflorous, but produce congested flower clusters at their stem tips.

COCOA TREE FRUIT

Cocoa plants (*Theobroma cacao*) produce their flowers and fruit on woody stems shaded by the foliage. The flowers are pollinated by midges, which prefer a dappled environment. Other cauliflorous trees include breadfruit (*Artocarpus altilis*), papaya (*Carica papaya*), calabash (*Crescentia cujete*), and many tropical figs (*Ficus* sp.). Outside of the tropics, rare examples of cauliflory among temperate trees and shrubs include the eastern redbud (*Cercis canadensis*), and the Judas tree (*C. siliquastrum*), which bear pink flowers on mature branches before the new leaves appear in spring.

THEOBROMA CACAO

Flower buds emerge from growing points, or meristems, located at nodes along the woody stems

The bold tufts of white stamens provide pollen for potential pollinators

Spent flowers will develop into vivid red, coiled pods

The stamens can grow up to 5 cm (2½ in) long

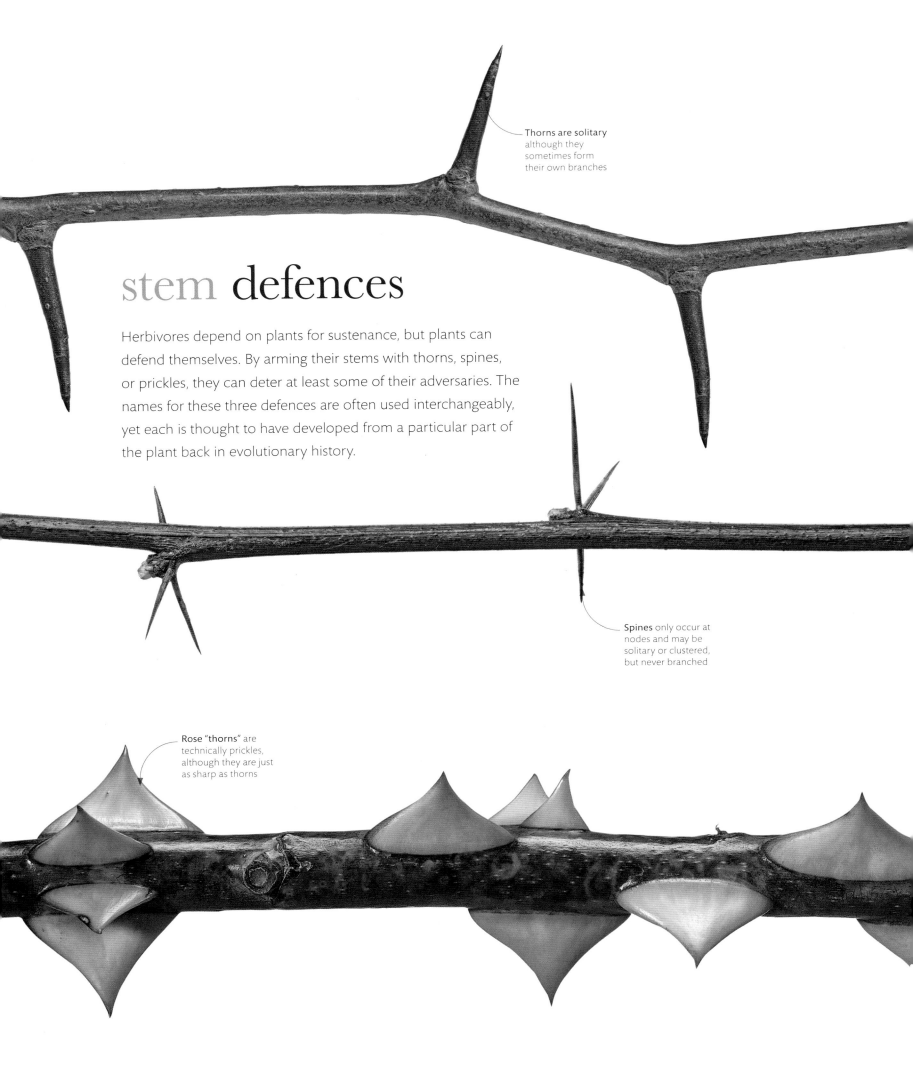

Thorns are solitary although they sometimes form their own branches

stem defences

Herbivores depend on plants for sustenance, but plants can defend themselves. By arming their stems with thorns, spines, or prickles, they can deter at least some of their adversaries. The names for these three defences are often used interchangeably, yet each is thought to have developed from a particular part of the plant back in evolutionary history.

Spines only occur at nodes and may be solitary or clustered, but never branched

Rose "thorns" are technically prickles, although they are just as sharp as thorns

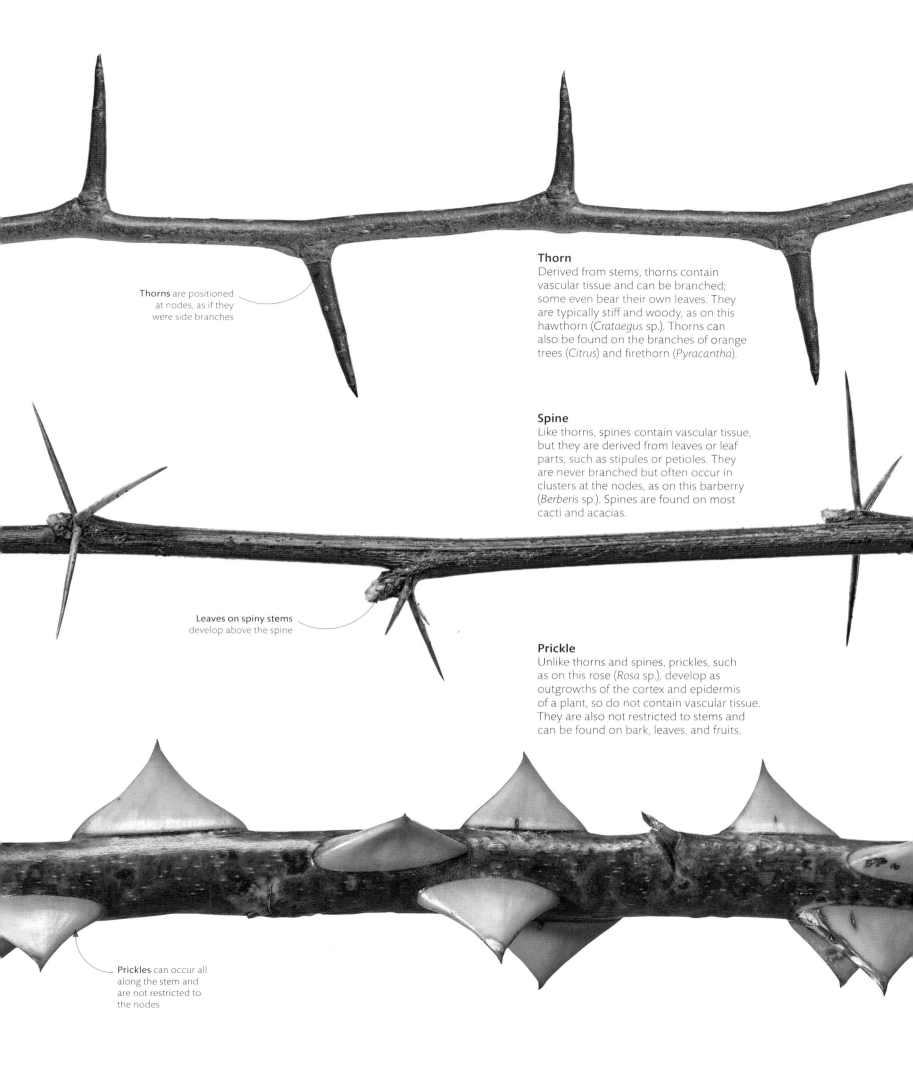

Thorn
Derived from stems, thorns contain vascular tissue and can be branched; some even bear their own leaves. They are typically stiff and woody, as on this hawthorn (*Crataegus* sp.). Thorns can also be found on the branches of orange trees (*Citrus*) and firethorn (*Pyracantha*).

Thorns are positioned at nodes, as if they were side branches

Spine
Like thorns, spines contain vascular tissue, but they are derived from leaves or leaf parts, such as stipules or petioles. They are never branched but often occur in clusters at the nodes, as on this barberry (*Berberis* sp.). Spines are found on most cacti and acacias.

Leaves on spiny stems develop above the spine

Prickle
Unlike thorns and spines, prickles, such as on this rose (*Rosa* sp.), develop as outgrowths of the cortex and epidermis of a plant, so do not contain vascular tissue. They are also not restricted to stems and can be found on bark, leaves, and fruits.

Prickles can occur all along the stem and are not restricted to the nodes

Ceiba pentandra

the kapok tree

Many rainforest trees grow to great heights and display huge buttress roots. Few, though, are more impressive than the majestic kapok. Given the right conditions, this important canopy species can grow as high as 70 m (230 ft), with buttresses that project up to 20 m (65 ft) from the trunk.

The deciduous kapok tree is found in the Americas, growing from southern Mexico to the southernmost limits of the Amazon Rainforest. It is also found in parts of West Africa. Exactly how the tree came to be native to both regions has been the subject of scientific inquiry. By analysing kapok DNA, experts now believe that the species made it to Africa after seeds were dispersed across the ocean from Brazil.

The kapok plays a vital role in the local ecology and culture wherever it grows. Its textured bark provides a habitat for bromeliads and other epiphytes, and also for reptiles, birds, and amphibians. The tree's ability to invade disturbed areas also makes it a key pioneer species, as it is one of the first trees to colonize open land after forest clearances.

The kapok's flowers open at night and emit a foul odour that attracts bats, its main pollinators. The tree is able to alter its pollination strategy according to the strength of the local bat population. In places where bats are plentiful, kapoks rely on them to spread pollen from tree to tree; where there are few bats however, the trees self-pollinate, guaranteeing at least some reproductive success each year.

After pollination, kapoks become laden with seedpods, each of which opens up to release about 200 seeds. The seeds are surrounded by cotton-like fibres that help them to disperse on the slightest breeze. Unopened seedpods float in water, so it is likely that the kapok initially travelled from the Americas to Africa by sea.

Prickly giant
A kapok's massive trunk can be up to 3 m (10 ft) in diameter. The huge prickles deter animals from chewing at its bark. The prickles eventually drop off as the tree ages.

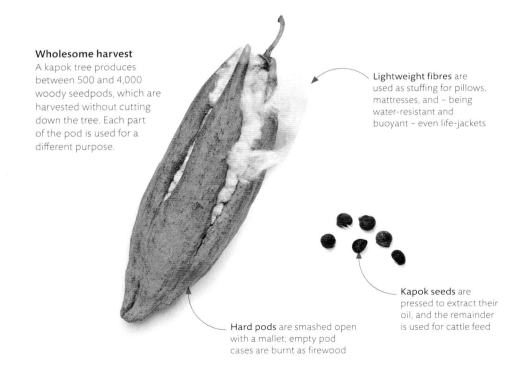

Wholesome harvest
A kapok tree produces between 500 and 4,000 woody seedpods, which are harvested without cutting down the tree. Each part of the pod is used for a different purpose.

Lightweight fibres are used as stuffing for pillows, mattresses, and – being water-resistant and buoyant – even life-jackets

Hard pods are smashed open with a mallet; empty pod cases are burnt as firewood

Kapok seeds are pressed to extract their oil, and the remainder is used for cattle feed

stems and resin

Trees are constantly under attack from a huge variety of insects, birds, fungi, and bacteria, which attempt to break through the bark and feed on the tissues below, either directly or through existing wounds. Many trees produce sticky resin to heal breaks in the bark and ensnare pests. Some resins even contain chemicals that attract predatory insects that will feed on the tree's attackers. Resin gradually hardens, and fossil resin, known as amber, often contains the remains of ancient insects.

Resin canals (or ducts) distribute resin around the stem or branch

CROSS-SECTION OF A PINE TRUNK

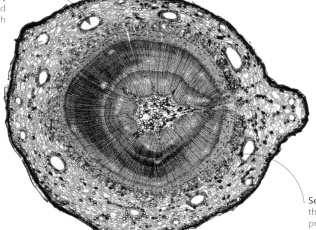

Releasing resin
Some trees only produce resin in response to damage, while others, such as this pine – stained to reveal the cell structure – develop resin ducts in their wood as a matter of course.

Sticky protection
The dripping resins produced by trees such as this oak are designed to heal wounds, whether inflicted by pests, or caused by physical damage from bad weather or fire. Plant resins are mixtures of organic compounds, that are used for many things – from perfumes and varnishes to adhesives. They are also the source of valuable commodities such as frankincense, turpentine, myrrh, and pitch.

Secretory cells surround the resin canals and produce the resin

Dracaena draco

the dragon tree

The dragon tree looks as though it belongs in a fantasy novel. This strange but beautiful monocot gets its name from the red resin, known as "dragon's blood", that it exudes when wounded. A member of the asparagus family, *Dracaena draco* has evolved a unique tree-like growth habit.

The dragon tree is endemic to parts of North Africa, the Canary Islands, Cape Verde, and Madeira. For the first few years of its life, the dragon tree is a single stem ending in a tuft of long, slender leaves. After 10 to 15 years' growth, the tree flowers for the first time. Long spikes full of fragrant white flowers erupt from the leaves, followed by bright red berries. A cluster of new buds then emerges from the top of the plant, and the buds develop into miniature versions of the original. They continue to grow for another 10 to 15 years before the branching process

Dragon grove
In the wild, the dragon tree grows in nutrient-poor soils. The thick trunk branches into upright arms ending in rosettes of lance-shaped, blue-green leaves up to 60 cm (2 ft) long. The species' status is listed as vulnerable.

begins again. Over time, this repeated branching gives the tree its unusual, umbrella-like shape. The lifespan of this species is thought to be about 300 years, but accurate ageing is difficult as the trees do not produce annual growth rings.

The tree's branches produce aerial roots that gradually snake their way down the trunk until they reach the soil. The roots emerge from wounds. If enough damage has been done to the tree, the roots can function as a new trunk, developing into a clone of the parent tree.

Once highly valued as a medicine and an embalming fluid, the dragon tree's blood-red resin is now used to stain and varnish wood. The resin is obtained by making cuts in the bark, but repeated wounding puts the tree at risk of infection. Dragon tree numbers in the wild are in decline, due to aggressive resin harvesting in the past and habitat loss today.

Dragon's blood oozes from wounds as a viscous liquid that dries and hardens

The deep-red resin has been used in dyes and traditional medicines since ancient times

Hardened resin
The tree bleeds a dramatically coloured resin called "dragon's blood", hence the species' name. The resin serves as a form of defence, deterring herbivores and keeping out pathogens.

storing food

Some plants have specially modified stems, roots, or leaf bases that live permanently underground. Swollen and densely packed with nutrients, these subterranean bulbs, corms, tubers, and rhizomes are dormant for part of the year, then sprout new shoots when the growing conditions are right. They are hidden from herbivores and can often spread underground to expand a plant's territory.

From bulb to bloom
Bulbous plants – which include familiar plants such as onions – have a short, squat stem called a basal plate, to which fleshy leaves (scales) are attached. The scales store the nutrients and water that the plant needs in order to produce flowers. These hyacinth bulbs flower in spring. Afterwards, the leaves photosynthesize to create more food, which is stored for the next year's flowering.

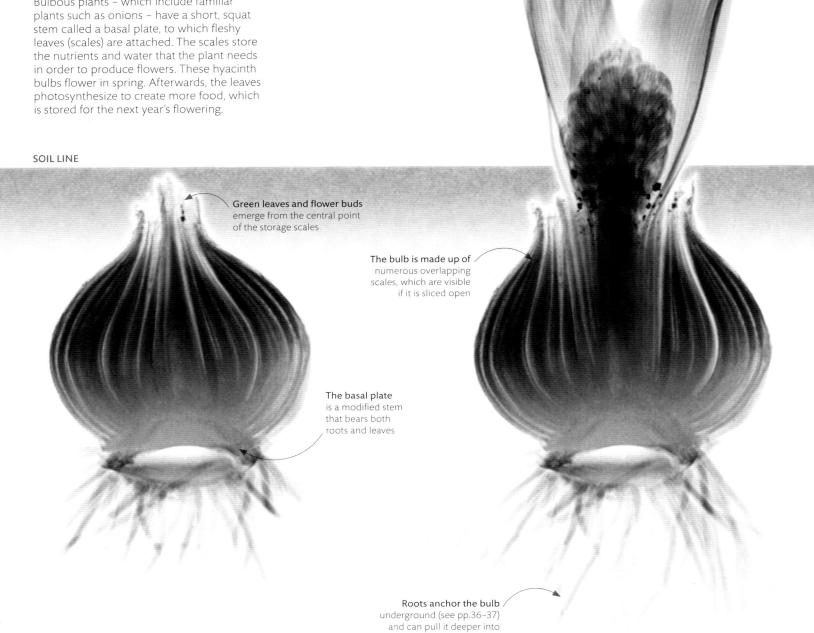

SOIL LINE

Green leaves and flower buds emerge from the central point of the storage scales

The bulb is made up of numerous overlapping scales, which are visible if it is sliced open

The basal plate is a modified stem that bears both roots and leaves

Roots anchor the bulb underground (see pp.36–37) and can pull it deeper into the soil if necessary

X-RAY OF A DORMANT HYACINTH BULB

EMERGING BULB WITH A DEVELOPING FLOWER HEAD

The fully expanded **leaves** photosynthesize to make carbohydrates, which are stored in the bulb scales below

The growing **leaves** surround and protect the delicate flower buds as they push up through the soil

The emerging **flower head** draws upon the energy reserves stored below. Bulbs with too little energy in reserve will not flower

New bulbs called **bulbils** form around the outer edge of the basal plate

BULB WITH MATURE LEAVES AND ABOUT TO FLOWER

STORAGE ORGANS

Bulbs are a round mass of swollen leaf bases that protect and feed new shoots. Corms and rhizomes are both a form of modified underground stem: a corm is bulb-shaped, whereas a rhizome grows horizontally just beneath or above the ground and produces shoots at its apex and along its length. Tubers can be derived from both stems and roots.

BULB

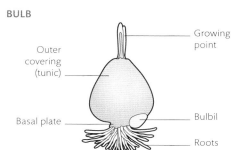

Outer covering (tunic)

Basal plate

Growing point

Bulbil

Roots

CORM

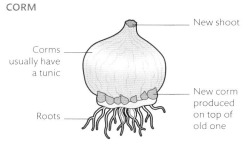

Corms usually have a tunic

Roots

New shoot

New corm produced on top of old one

RHIZOME

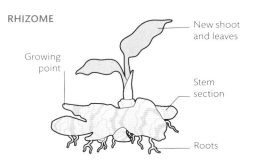

Growing point

New shoot and leaves

Stem section

Roots

TUBER

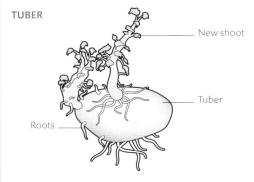

New shoot

Tuber

Roots

HOMES FOR ANTS

The living quarters that ant plants provide for ants are known as domatia and can develop in different parts of the plant. *Dischidia* vines house ants in swollen leaves, whereas *Lecanopteris* ferns shelter them within their rhizomes. *Myrmecophila* orchids accommodate ants in hollow, bulb-like swellings of the stem, and some acacias host them within hollow spines. *Myrmecodia* and *Hydnophytum* both have swollen stem tubers with intricate internal structures, giving ants chambers for various uses.

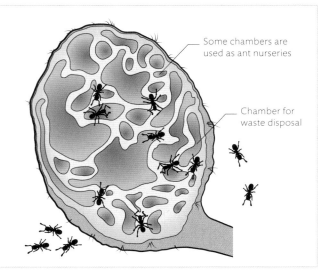

Some chambers are used as ant nurseries

Chamber for waste disposal

DOMATIA IN A *HYDNOPHYTUM* TUBER

Good tenants

The elaborate stem tubers of the ant-house plant (*Myrmecodia beccarii*) include rough-walled cavities where ants deposit their waste, as well as remains of their prey and corpses. Nodules on the walls of the cavities absorb nutrients from this slurry, providing this tree-dwelling plant with vital elements that it would otherwise find hard to access. The plant shown right, from Australia and grown at Kew in 1888, is depicted in an illustration from *Curtis's Botanical Magazine*.

Partners in crime

Some ant plants often grow together. Here, the brown stem tuber of *Hydnophytum* is nestled within the yellow leaves of a *Dischidia* species. Both plants provide a home for ants.

beneficial
relationships

Insects can be a benefit or a burden for plants. Although some provide their services as pollinators, others are ruthless leaf-eaters that weaken the plant to satisfy their appetites. A handful of plants, however, have struck up a mutually beneficial (symbiotic) relationship with ants. So-called "ant plants" provide a secure home for the ant colony and, in return, the ants protect the plant by attacking anything that comes near. As many ant plants are epiphytes, unconnected to the soil and its nutrients, the rich fertilizer formed from the ants' waste is also a crucial food supply.

STEM STRUCTURES

A tree fern trunk is really an upright rhizome supported by a dense, encircling mass of roots and fibres. Banana stems are not really stems at all, but overlapping layers of leaf sheaths; the true "stem" is a rhizome (see p.86) that is concealed underground.

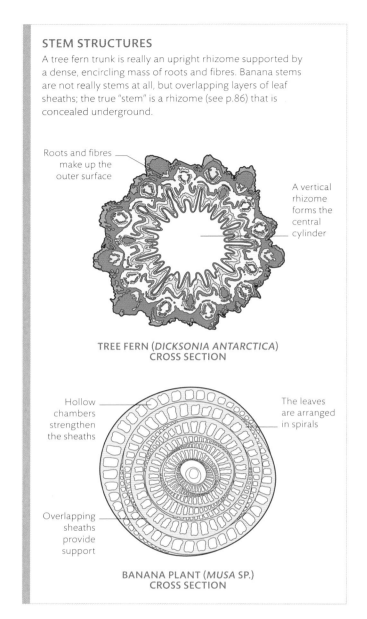

Roots and fibres make up the outer surface

A vertical rhizome forms the central cylinder

TREE FERN (*DICKSONIA ANTARCTICA*)
CROSS SECTION

Hollow chambers strengthen the sheaths

The leaves are arranged in spirals

Overlapping sheaths provide support

BANANA PLANT (*MUSA* SP.)
CROSS SECTION

Tree ferns

Rhizomes are a type of swollen stem that acts as a food store for the plant. They usually grows horizontally, but in tree ferns they grow upwards. Masses of roots and fibres emerge from the rhizome and grow around it, forming a thick, protective mantle that supports it in an upright position.

fibrous trunks

Not all trees are true trees. Well-known species such as pines and firs (conifers), or oaks and maples (deciduous), have woody trunks with characteristic growth rings and an outer layer of bark. Tree ferns and banana plants, however, have quite a different stem structure and do not produce wood or bark. Their sturdy, upright stems are supported by densely packed fibres, roots, or tightly packed, overlapping sheaths of leaves.

Plants of the Coromandel Coast
William Roxburgh's *Plants of the Coast of Coromandel* (1795), which included this illustration of the toddy palm (*Borassus flabellifer*), was published under the direction of Sir Joseph Banks, the long-time president of the Royal Society.

Company style
Painted by an unknown artist, this watercolour illustration of a fan palm (*Livistona mauritiana*) is attributed to the Company School, a group of Indian artists who worked under the patronage of the East India Company, using the distinctive Indo-European Company style.

west meets east

During the 18th and 19th century, with the expansion of British influence into India, scientists and natural historians engaged by the East India Company began to explore and document the richness and diversity of the country's flora. Highlights of the resulting body of work were striking illustrations that display a unique fusion of Western science and Eastern art.

William Roxburgh (1751–1815), often referred to as the father of Indian botany, was the director of the Calcutta Botanic Garden when he began to commission local artists to create the botanical illustrations used in his important books. More than 2,500 life-sized paintings would feature in his landmark *Flora Indica*.

Influenced by the miniature painters of the 16th- and 17th-century Mughal Empire, the Indian artists created a style that combined the precise detail of Western botanical illustration with the decorative approach and immediacy of their own art. This hybrid style of artwork was well suited to botanical illustration. Working with the naturalist Johan König, who had been a pupil of the great Swedish taxonomist Carl Linnaeus, Roxburgh commissioned many artworks that would become "type" drawings identifying particular wild species.

Characteristic arrangement
Painted in watercolour on paper, this hand-painted copy of an illustration commissioned by William Roxburgh depicts the Indian redwood (*Caesalpinia sappan*). The way that the plant has been cropped, so that parts of it are not visible, is characteristic of this style of painting. The illustration is inscribed "Received from Rodney, 9th June 1791". This refers to the East India Company merchant ship that delivered artworks from India to London.

> " The artworks embody essential qualities of their diverse patronage. "

PHYLIS I. EDWARDS, *INDIAN BOTANICAL PAINTINGS*, 1980

HOW CLIMBERS FIND SUPPORTS

Plants cannot see, so they have to find another way to locate a support. Some vines detect shade and grow towards it, as that is likely to lead them to the foot of a tree. Others can follow trails of chemicals that lead to a suitable host, and avoid those that lead to other vines. Young stems revolve as they grow, which can help them snag a neighbouring branch. Once in position, twining stems or tendrils hook on to the support.

Stem tendril

Flexible stem

Bamboo support

Emerging flowers have the potential to release many seeds, further expanding the plant's territory

Long leaf stalks allow the leaves to reach away from the support and face the sun

Bindweed stems can quickly choke out neighbouring plants, reaching over 3 m (10 ft) in length each year

To expose their leaves to light, vines climb up other plants and escape the shade

Takeover by twining

Bemoaned by many gardeners, the twining stems of hedge bindweed (*Calystegia sepium*) are one of the secrets of its success. Growing through shrubs and perennials, it quickly smothers their foliage with its own as it successfully competes with them for sunlight. Underground, the plant is equally vigorous, spreading its white rhizomes in all directions.

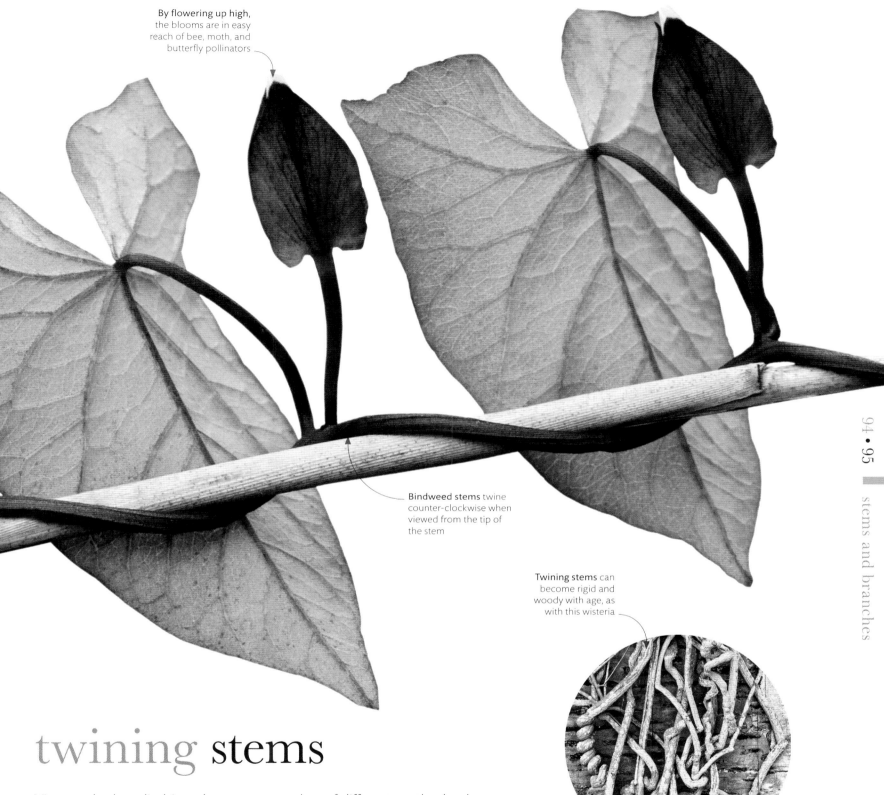

By flowering up high, the blooms are in easy reach of bee, moth, and butterfly pollinators

Bindweed stems twine counter-clockwise when viewed from the tip of the stem

Twining stems can become rigid and woody with age, as with this wisteria

twining stems

Vines and other climbing plants use a number of different methods when scrambling up a support. Tendrils, aerial roots, and hooked prickles all attach themselves to supports, but in some climbers, the stem itself can cling by twining. Some twining plants twist clockwise, while others twist anti-clockwise, and this distinction, which may have a genetic basis, can be used to tell some species of climbers apart. Beans and bindweed twine anti-clockwise, while hops and honeysuckle twine clockwise.

Touch and feel
The ability of a stem to twine around another is due to thigmotropism. When climbing stems and tendrils detect the presence of a support, one side of the growing point begins to grow faster than the other, making the stem bend.

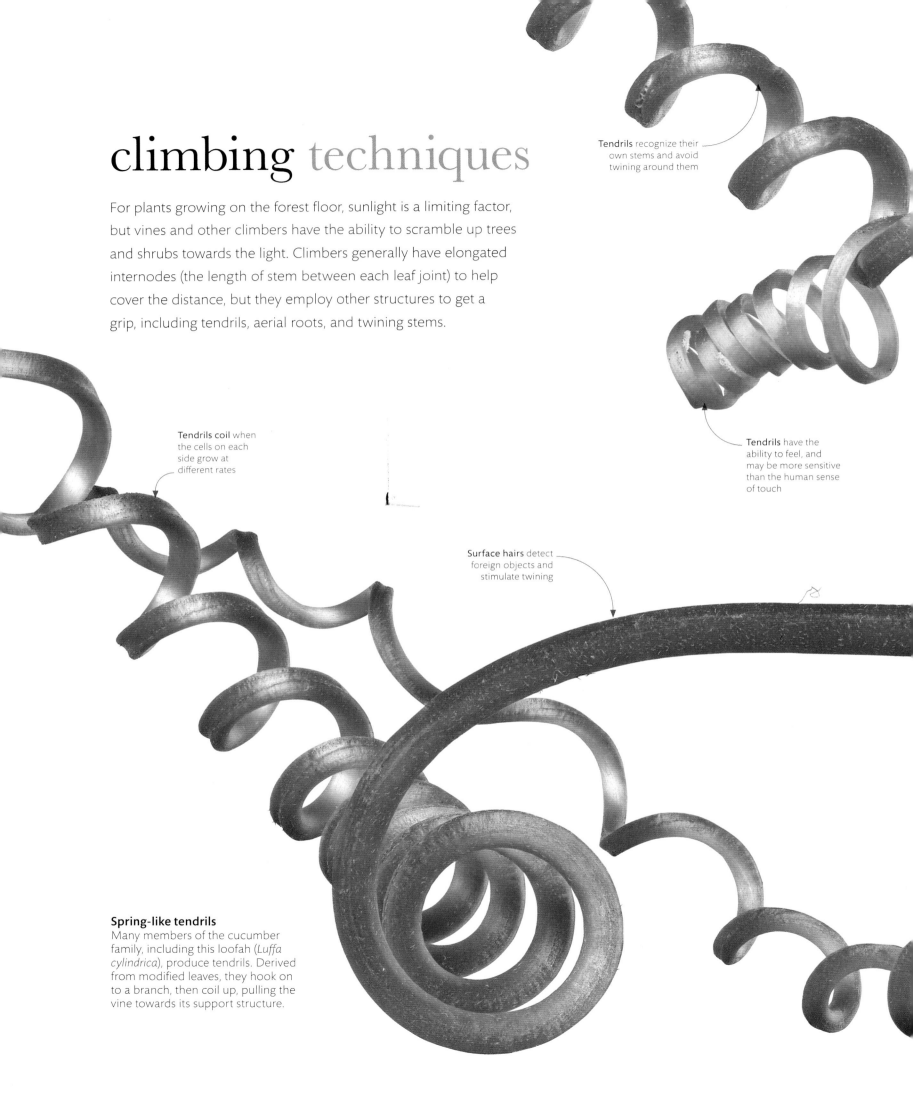

climbing techniques

For plants growing on the forest floor, sunlight is a limiting factor, but vines and other climbers have the ability to scramble up trees and shrubs towards the light. Climbers generally have elongated internodes (the length of stem between each leaf joint) to help cover the distance, but they employ other structures to get a grip, including tendrils, aerial roots, and twining stems.

Tendrils recognize their own stems and avoid twining around them

Tendrils coil when the cells on each side grow at different rates

Tendrils have the ability to feel, and may be more sensitive than the human sense of touch

Surface hairs detect foreign objects and stimulate twining

Spring-like tendrils
Many members of the cucumber family, including this loofah (*Luffa cylindrica*), produce tendrils. Derived from modified leaves, they hook on to a branch, then coil up, pulling the vine towards its support structure.

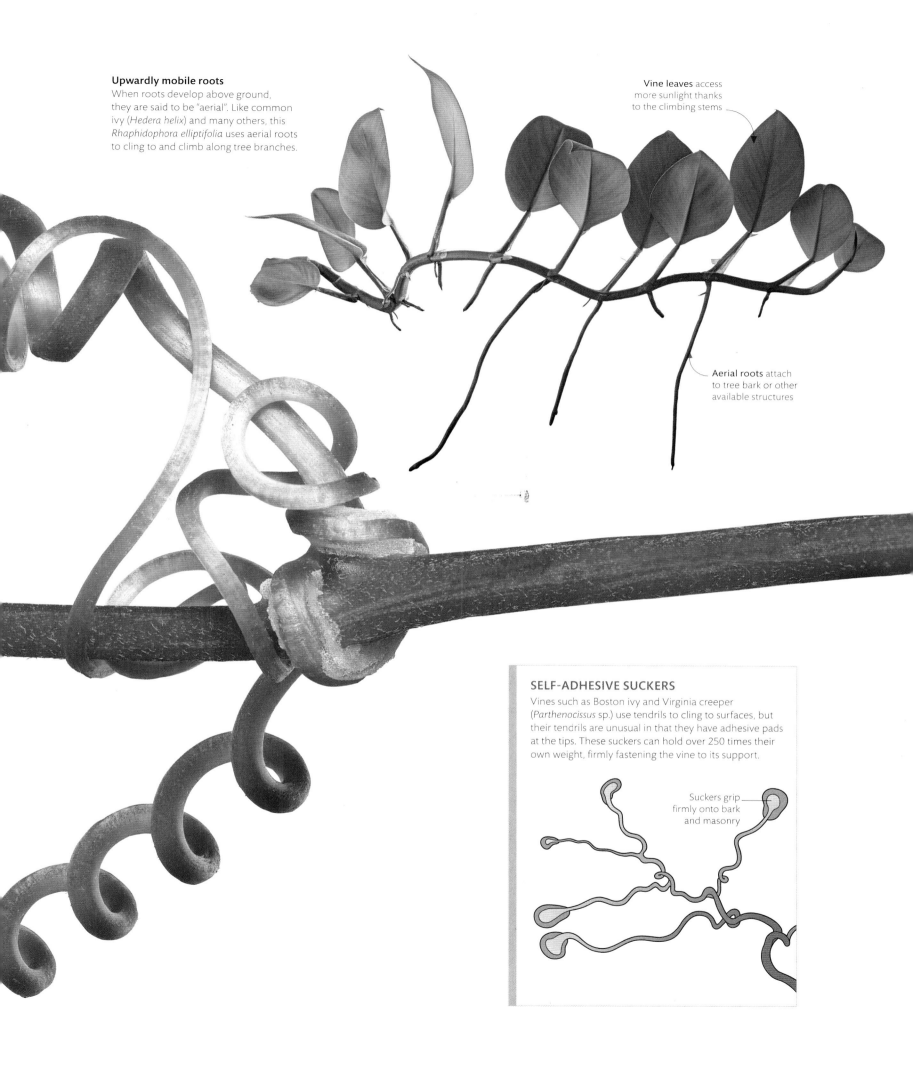

Upwardly mobile roots
When roots develop above ground, they are said to be "aerial". Like common ivy (*Hedera helix*) and many others, this *Rhaphidophora elliptifolia* uses aerial roots to cling to and climb along tree branches.

Vine leaves access more sunlight thanks to the climbing stems

Aerial roots attach to tree bark or other available structures

SELF-ADHESIVE SUCKERS

Vines such as Boston ivy and Virginia creeper (*Parthenocissus* sp.) use tendrils to cling to surfaces, but their tendrils are unusual in that they have adhesive pads at the tips. These suckers can hold over 250 times their own weight, firmly fastening the vine to its support.

Suckers grip firmly onto bark and masonry

SPINES AND WATER CONSERVATION

SPINES AND WATER CONSERVATION

Cactus spines are derived from leaves and protect their succulent stems from the attentions of thirsty animals. Spines also intercept water and direct it down to the ground; they also shade the plant from sunlight and slow the passage of air around the plant. Both strategies reduce water loss.

CROSS SECTION OF A TYPICAL CACTUS STEM

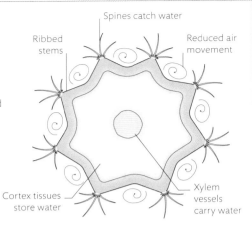

Spines catch water

Ribbed stems

Reduced air movement

Cortex tissues store water

Xylem vessels carry water

Water storage

The most famous members of the cactus family are giants like the saguaro cactus (*Carnegiea gigantea*, see p.100), but tiny treasures like this *Mammillaria infernillensis* are also well adapted to drought. Its thick skin has a waxy coating, greatly reducing water loss.

storage stems

Cacti are renowned for their ability to hoard water in their succulent stems. Most species live in arid areas where rainfall is scarce, so when the heavens open, they must take advantage of the deluge. To absorb as much water as possible, and quickly, many cacti have ribbed stems that expand like an accordion. This prevents them from splitting as they become rehydrated.

Lost leaves

Prickly pears such as *Opuntia phaeacantha* produce tiny leaves on new stem segments, but they are shed quickly to conserve water.

Individual stem segments can be dropped during extended droughts

Large spines are surrounded by tiny, hair-like spines called glochids, which detach and irritate animal skin

Tufts of white hair reduce water evaporation and reflect sunlight, cooling the cactus down

Sharp spines take the place of leaves, in order to protect the plant from predation

Carnegiea gigantea

the saguaro cactus

What Western film would be complete without a backdrop of giant saguaro cacti? These towering succulents are icons of the Sonoran Desert in Arizona, California, and northwest Mexico, and a special community of bats, birds, and other animals has evolved around them.

By far the most impressive aspect of the saguaro cactus is its size. Individuals regularly reach 15 m (50 ft) or more in height and can weigh as much as 2,000 kg (4,400 lb). Much of that bulk consists of stored water – a precious commodity in the desert. On the rare occasions when it rains, the ribs that run the length of the saguaro's stem can expand to allow the plant to swell and take up as much water as possible through its extensive, shallow root system. Once stored, the water needs protecting: the saguaro is covered with spines and bristles that not only deter herbivores from feeding on its succulent tissues, but also create shade and reduce airflow close to the saguaro's skin, minimizing water loss to the air.

In spring, saguaro cacti usually put on an impressive floral display. Dense clusters of bright white flowers appear from the apex of the main stem and arms.

The flowers are pollinated by birds and insects by day, and by bats at night. The nectar in the saguaro's blooms contains compounds that help female lesser long-nosed bats produce enough milk to feed their young. After the flowers come the fruits, providing an energy-rich food for a wide range of desert animals.

The saguaro has a particularly close relationship with the Gila woodpecker. This bird excavates a hole in the cactus in which it can build its nest. Vacated woodpecker holes are subsequently used as shelter and nesting sites by many other species of bird, mammal, and reptile.

Saguaro sentinels
Saguaro cacti stand like alert guards watching over the desert. Slow-growing but long-lived, saguaros can survive for 200 years or more. They only grow arms after they reach 50 to 100 years old.

Crested saguaro
Occasionally saguaro cacti take on a crested form. This fan-like appearance is due to changes in the growing tip (apical meristem). The cause is not known: it may result from a genetic mutation, or from physical damage by lightning or frost.

New arms can continue to sprout on the crest

Fan-like crests are thought to develop on only about one in every 200,000 cacti

leafless stems

Leaves produce food for plants, but moisture quickly evaporates through their large surface area. In the dry, harsh climate of the desert, some plants have adapted by not growing leaves. Instead, their green succulent stems take on photosynthesis. By taking in carbon dioxide at night and storing it in their stems, these plants can photosynthesize with pores closed during the hot sunlight hours.

Succulent stems only open their pores to allow gas exchange at night

Stems at work

Many succulents, such as this *Euphorbia woodii*, a South African member of the spurge family, show the same adaptation to their environment as cacti. They do not have spines, but their very reduced leaves mean that they rely on their succulent stems to carry out photosynthesis and produce the carbohydrates that all plants need in order to grow.

A poisonous sap in the stems of this succulent *Euphorbia* deters herbivores

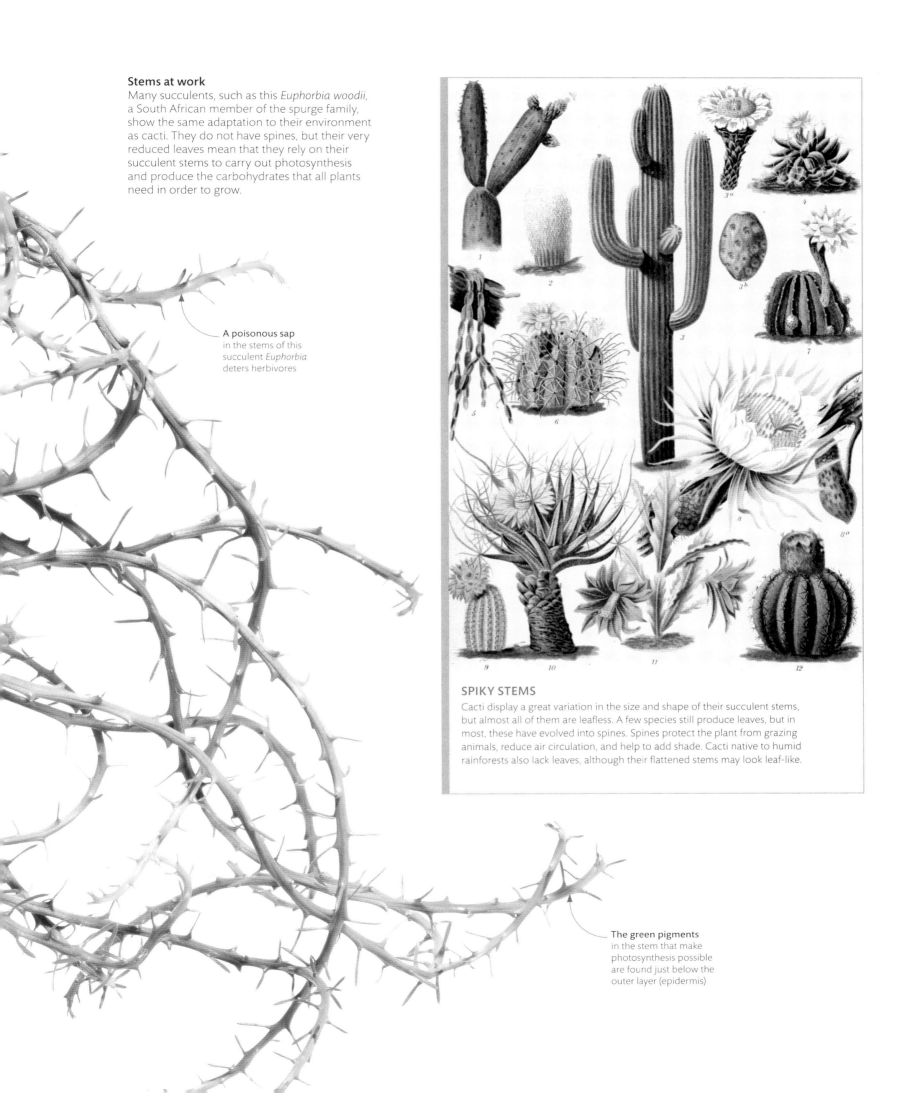

SPIKY STEMS

Cacti display a great variation in the size and shape of their succulent stems, but almost all of them are leafless. A few species still produce leaves, but in most, these have evolved into spines. Spines protect the plant from grazing animals, reduce air circulation, and help to add shade. Cacti native to humid rainforests also lack leaves, although their flattened stems may look leaf-like.

The green pigments in the stem that make photosynthesis possible are found just below the outer layer (epidermis)

Strawberry runners
A number of species colonize new territory using stolons. They include the edible strawberry (*Fragaria* x *ananassa*), spider plant (*Chlorophytum comosum*), and mock strawberry (*Duchesnea indica*, syn. *Fragaria indica*) – shown here in a drawing made in India for Robert Wright in 1846. Their stems produce new plants at the nodes, and become independent once severed from the parent.

new plants from stems

Stems make it possible for a plant to expand its range in several ways. The ground-hugging or prostrate stems of some plants can root as they spread, while subterranean rhizomes do the same underground. Some plants that are otherwise upright produce slender, elongated horizontal stems that creep across the soil or just beneath it, and form baby plants at the nodes; these stems are called stolons or runners.

STEMS UNDERGROUND
Many plants develop rhizomes, corms, or tubers, which are derived from stems. These underground "stems" grow on or just below the soil surface, and they not only allow plants to survive adverse conditions, but they also provide a means of propagation. Any pieces broken off the parent can root and form new plants. *Crocosmia* corms can also produce stolons that distribute new corms a short distance away from the parent.

CROCOSMIA **CORM OFFSETS**

Crocosmia flowers on long, upright stems

New corms form on top of old corms

Dryadeæ

Rungiah, del

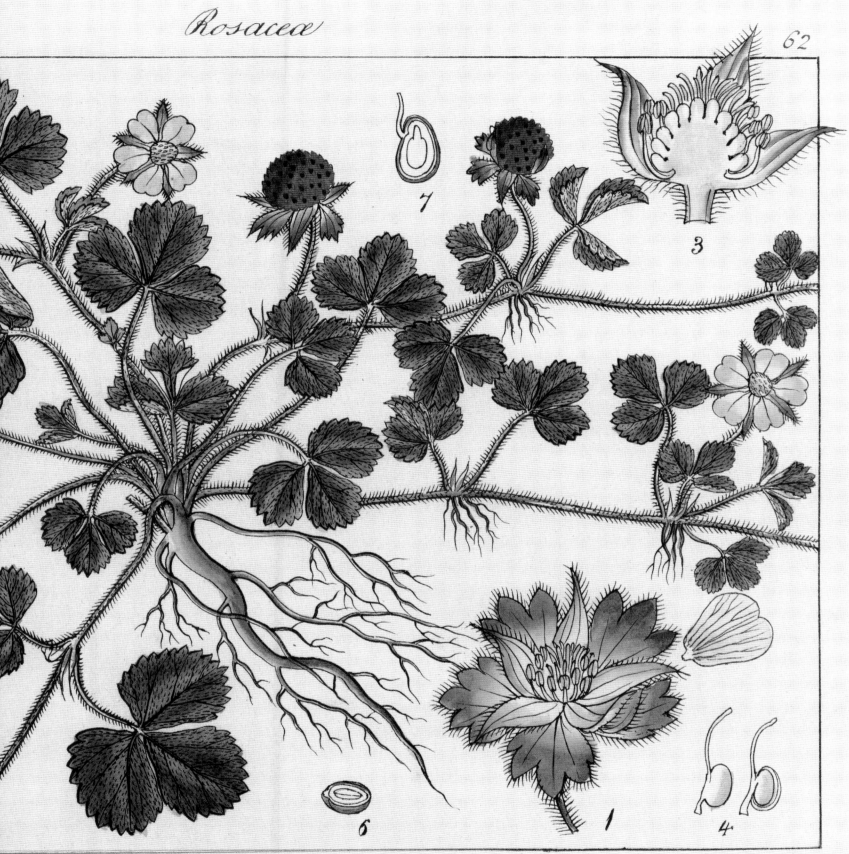

7

3

6

1

4

Dumphy, Lith.

Fragaria indica (Andr)

moso bamboo

The canopy of a grove of moso bamboo can reach a level of up to 30 m (100 ft) above the ground, making it easy to think these giant bamboos are trees. But moso bamboo is a grass, albeit a very tall, woody one. Like other grasses, it is characterized by a jointed stem called a culm.

Although it is synonymous with Japanese culture, moso bamboo (*Phyllostachys edulis*) is actually native to mountain slopes in warm, temperate regions of China, and only naturalized to Japan. This species is of great economic importance throughout Asia, where it is used as a food source, a building material, and a fibre for textiles and making paper.

Moso bamboo has an astonishing rate of growth, and new stems are able to add more than 1 m (3 ft) to their height each day. Growth is also vigorous below the surface of the soil: dense mats of roots and rhizomes (underground stems) spread out relentlessly, throwing up new shoots to colonize territory. Such vegetative growth is the plant's main reproductive strategy. As a result, entire hillsides can be made up of clones of a single individual. While it can also reproduce sexually, this species flowers only once every 50 to 60 years.

When it does flower, however, it produces many thousands of seeds that are quick to germinate.

Moso's aggressive growth habit is cause for concern when it is introduced outside its native range. Individuals can quickly escape gardens and invade surrounding areas. Producing impenetrable root mats, heavy leaf litter, and dense shade, they easily smother other plant species.

Young moso shoots are edible, but like many bamboos, moso protects itself with a potent chemical cocktail that includes oxalic acid and cyanide compounds. With adequate boiling, the compounds break down and render the shoots safe to eat.

Bamboo canopy, Japan
Covering 16 sq km (6 sq miles), the Sagano Bamboo Forest near Kyoto is treasured for the beauty and tranquility of its dense bamboo groves.

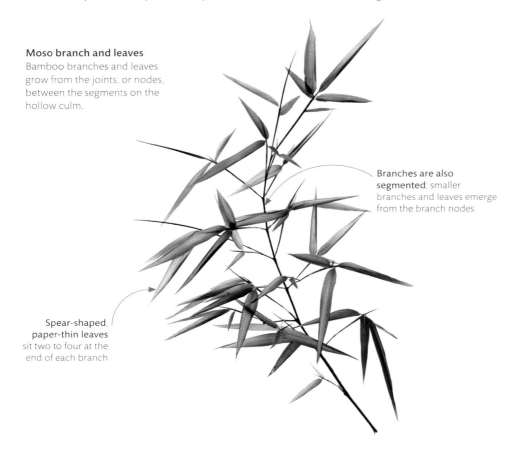

Moso branch and leaves
Bamboo branches and leaves grow from the joints, or nodes, between the segments on the hollow culm.

Branches are also segmented; smaller branches and leaves emerge from the branch nodes

Spear-shaped, paper-thin leaves sit two to four at the end of each branch

leaves

leaf. a flat, usually green, structure, attached to a plant stem directly or by a stalk, in which photosynthesis and transpiration take place.

Deciduous broadleaves
Many plants, such as this sugar maple (*Acer saccharum* subsp. *saccharum*), have large, flat leaves, maximizing the surface area available for photosynthesis.

types of leaf

Leaves fall into two main categories: evergreens, which stay on the plant all year, or deciduous, which are discarded seasonally. Leaves are a significant investment for a plant. Evergreen leaves may minimize the demands on its resources in the long term but, if leaves are unlikely to survive year-round due to unfavourable conditions, deciduous leaves may be more beneficial.

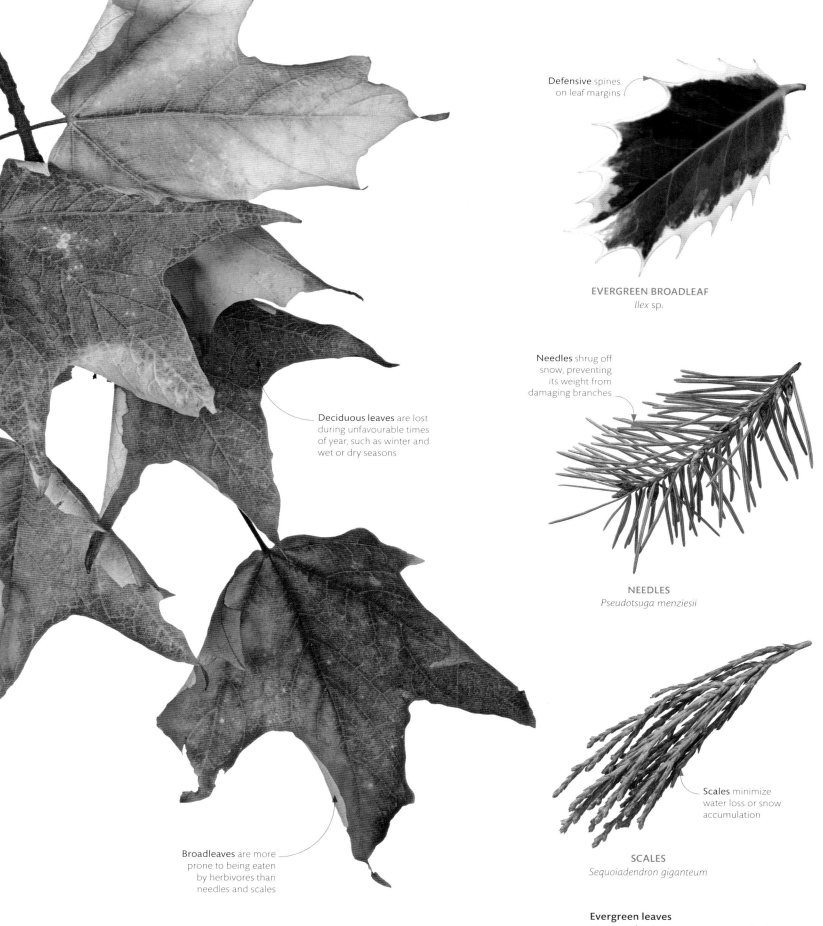

Defensive spines
on leaf margins

EVERGREEN BROADLEAF
Ilex sp.

Needles shrug off
snow, preventing
its weight from
damaging branches

NEEDLES
Pseudotsuga menziesii

Deciduous leaves are lost
during unfavourable times
of year, such as winter and
wet or dry seasons

Scales minimize
water loss or snow
accumulation

SCALES
Sequoiadendron giganteum

Broadleaves are more
prone to being eaten
by herbivores than
needles and scales

Evergreen leaves

Most conifers have evergreen leaves that are
reduced to needles or scales. These have a
smaller surface for photosynthesis, but can
photosynthesize for more of the year. They
are also adapted to survive cold winters.

leaf structure

Most leaves are full of cells that harvest light for photosynthesis. These mesophyll cells are supplied with water and nutrients by a network of veins. The veins also carry away carbohydrates made during photosynthesis to other parts of the plant. Pores on the leaf surface – stomata – open to take in carbon dioxide and close to avoid water loss. Evaporation from the rest of the leaf surface is prevented by a waterproof waxy coating called the cuticle.

Inside foliage
Taro leaves (*Colocasia esculenta*) are covered by a single layer of cells called the epidermis (blue). The interior has palisade mesophyll (green) and spongy mesophyll (yellow) cells. The grey structure is a vein, or vascular bundle, made up of xylem and phloem vessels.

Lateral veins branch off the midrib to carry water and nutrients to all areas of the leaf

Waxy cuticle prevents water loss

Xylem transports nutrients and water from the root to the shoot

Phloem carries carbohydrates made during photosynthesis to other parts of the plant

The leaf stem extends along the centre of the leaf, where it is known as the midrib

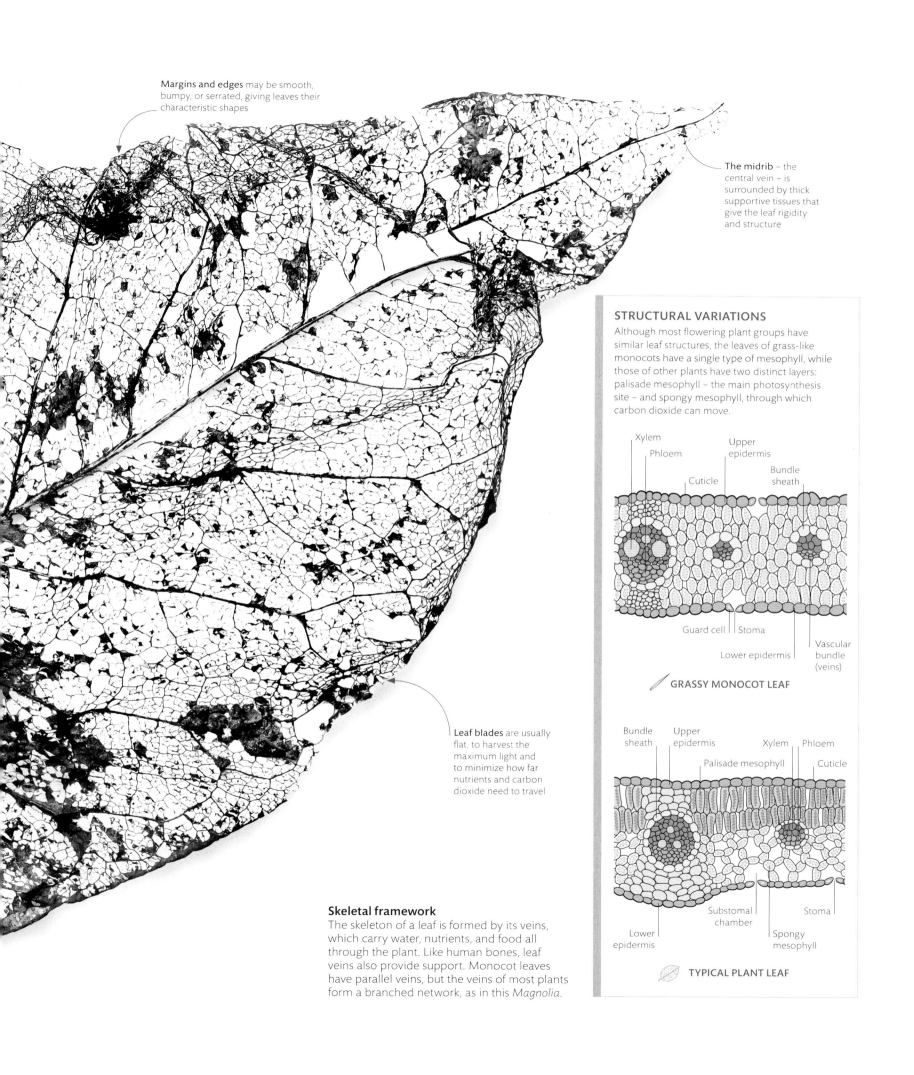

Margins and edges may be smooth, bumpy, or serrated, giving leaves their characteristic shapes

The midrib – the central vein – is surrounded by thick supportive tissues that give the leaf rigidity and structure

Leaf blades are usually flat, to harvest the maximum light and to minimize how far nutrients and carbon dioxide need to travel

STRUCTURAL VARIATIONS

Although most flowering plant groups have similar leaf structures, the leaves of grass-like monocots have a single type of mesophyll, while those of other plants have two distinct layers: palisade mesophyll – the main photosynthesis site – and spongy mesophyll, through which carbon dioxide can move.

Xylem

Phloem

Cuticle

Upper epidermis

Bundle sheath

Guard cell   Stoma

Lower epidermis

Vascular bundle (veins)

GRASSY MONOCOT LEAF

Bundle sheath

Upper epidermis

Palisade mesophyll

Xylem   Phloem

Cuticle

Lower epidermis

Substomal chamber

Spongy mesophyll

Stoma

TYPICAL PLANT LEAF

Skeletal framework

The skeleton of a leaf is formed by its veins, which carry water, nutrients, and food all through the plant. Like human bones, leaf veins also provide support. Monocot leaves have parallel veins, but the veins of most plants form a branched network, as in this *Magnolia*.

compound leaves

Compound leaves are divided into two or more leaflets, providing plants with more photosynthetic surface area for the same investment in resource-hungry support tissues, such as branches. When leaflets arise from a single common point, the leaves are called palmate. When leaflets arise from several locations along the central stalk, or rachis, the leaf is said to be pinnate.

TRIPINNATE (ALTERNATE)
Pteridium sp.

Pinna

Rachis

Segment

Pinnule

Pinnae are arranged
in a staggered fashion
along the rachis

Fern fronds
Most ferns, such as this *Pteridium* leaf, exhibit varying degrees of pinnation. The leaflets are referred to as pinnae, and these are often further subdivided into pinnules. Pinnules in turn are divided into segments.

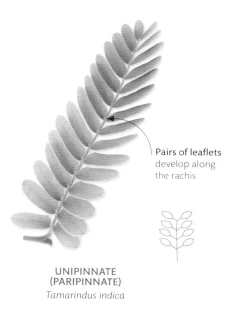

Pairs of leaflets develop along the rachis

UNIPINNATE (PARIPINNATE)
Tamarindus indica

Terminal leaflet forms at the end of the rachis

UNIPINNATE (IMPARIPINNATE)
Carya ovata

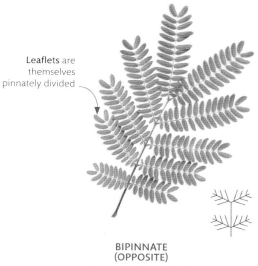

Leaflets are themselves pinnately divided

BIPINNATE (OPPOSITE)
Leucaena leucocephala

Winged petiole (leaf stalk) resembles the true leaf

UNIFOLIATE
Citrus hystrix

Two leaflets grow off a single petiole

BIFOLIATE
Hymenaea courbaril

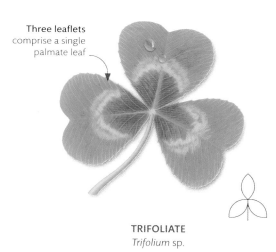

Three leaflets comprise a single palmate leaf

TRIFOLIATE
Trifolium sp.

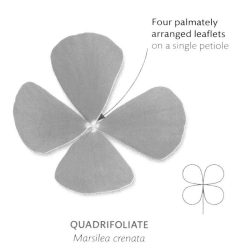

Four palmately arranged leaflets on a single petiole

QUADRIFOLIATE
Marsilea crenata

Palmately compound leaves with five leaflets

FIVE-PALMATE
Aesculus pavia

Variable numbers of leaflets form, depending on the growing conditions

MULTIFOLIATE
Cannabis sp.

Acanthus wallpaper design, 1875
Morris's wallpapers and fabrics featured stylized designs based on large, repeating patterns of flowers, leaves, or fruit. In this, his most expensively produced wallpaper, he used the deeply cut leaves of the acanthus, a plant that has appeared in architecture and art since ancient times. Acanthus was printed by the London firm Jeffrey & Co., using 15 natural dyes and 30 separate wood blocks for each full repeat.

66 ... any decoration is futile... when it does not remind you of something beyond itself. 99

WILLIAM MORRIS, *LECTURE ON PATTERN*, 1881

Art of glass
This decorative stained-glass window design is by the Italian artist Giovanni Beltrami (1860–1926). The delicate leaf and floral motifs are typical of the Art Nouveau style.

plants in art

designed by nature

The Arts and Crafts movement of the late 19th century developed as a reaction to the effects of industrialization on the lives of ordinary people and to the shoddy quality and design of mass-produced goods. Nostalgia for the simple ways of the past, fine materials, and honest craftsmanship was at the core of the movement, and its leading craftsmen and designers were primarily inspired by the forms of the natural world.

Horse chestnut, 1901
Scottish artist Jeannie Foord's botanic pantings were composed with a designer's eye. Their homage to the simple, naturalistic beauty of everyday leaves and flowers typifies the values of the Arts and Crafts movement.

The driving force behind the Arts and Crafts movement was the British craftsman William Morris. Virtually all of his designs for wallpapers and textiles show twining tendrils, leaves, and flowers. They were named after the plants they featured, but Morris's designs were stylized evocations of their forms, rather than botanically accurate reproductions.

Morris's study of ancient herbals. medieval woodcuts, tapestries, and illuminated manuscripts informed his designs, and he revived traditional crafts such as woodblock printing and hand weaving. He urged design students to correct their "mannered work" with a diligent study of nature, study of the different ages of art, and imagination.

Influenced in part by the Arts and Crafts movement, Art Nouveau artists and designers regarded nature as the underlying force of life and developed their own distinctive vocabulary of motifs based on the organic shapes of swirling plant roots, tendrils, and flowers, often fused with images of sensuous women.

developing leaves

Like all parts of a plant, leaves develop from clusters of dividing cells. Many species of tree, notably conifers, produce leaves continually, but deciduous broadleaf trees can only produce them at specific times of the year. In autumn, deciduous trees produce hardened, dormant buds that contain partially developed leaves. The buds are protected over winter and establish new leaves quickly in spring when the risk of frost is over.

As the bud begins to swell, the scales enlarge, to carry on protecting the young leaves

Modified leaves called scales protect the buds

Sycamore leaf buds are produced in pairs on opposite sides of the stem

The folds of the leaf within the bud dictate its final shape

Leaves are formed in pairs within each bud, pushing against each other to grow into the correct shape

Within the bud, the blade (lamina) of the leaf is folded along the veins

Newly burst leaves are still relatively crinkled

Bud scales eventually fall off as the leaves grow

The emerging leaves slowly expand and flatten out

Red pigments protect young sycamore leaves from being damaged by light

Young leaf tissues are soft and easily damaged

Timing it right

New leaves have to emerge at just the right time; if the buds burst too early, the young leaves risk being damaged by frost; if they emerge too late, they lose valuable growing time. Trees such as this sycamore (*Acer pseudoplatanus*) monitor the number of cold days to estimate when winter is over, and wait for warmer temperatures before the buds start bursting.

WHY LEAVES ARE GREEN

Leaves contain chlorophyll, a pigment that harvests light energy for use in photosynthesis (see pp.128–129). Chlorophyll is stored on stacks of membranes within tiny particles in the leaf cells called chloroplasts. Sunlight contains every colour, and chlorophyll absorbs them all except for green. The green light reflects out of the leaf, or passes through it, and this makes the leaf look green.

Light

Reflected light

Chloroplast

Transmitted light

CHLOROPLAST ABSORBING COLOURS FROM LIGHT

Expanded sycamore leaves have five palmate lobes

SPORE-BEARING FRONDS

Ferns do not flower but produce spores (see p.338) in speckles on their leaf undersides. Each speckle, or sorus, has a protective cover – the indusium, shown as a semi-circular structure on the drawing below – which shrivels to enable the fern to release its spores.

DRYOPTERIS FILIX-MAS

unfurling ferns

Developing fern fronds are tightly curled into structures called fiddleheads, which protect the delicate growing tips at their centre. As they slowly unfurl, the lower parts of the leaf toughen and begin to photosynthesize, providing the energy that powers the development of the rest of the leaf. This uncoiling process, known as circinate vernation, is mainly found in ferns and palm-like cycads.

Why ferns are furled
Ferns produce relatively few large leaves, but each represents a major investment of resources. The furled leaves only have a limited ability to photosynthesize, but remain shielded from herbivores. The loss of photosynthesis is outweighed by the reduction in damage from grazing insects.

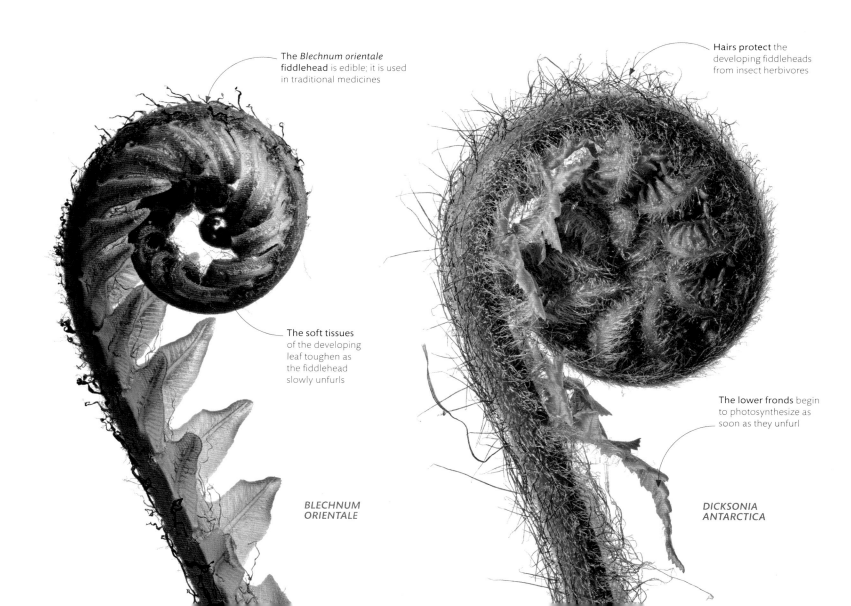

The *Blechnum orientale* fiddlehead is edible; it is used in traditional medicines

Hairs protect the developing fiddleheads from insect herbivores

The soft tissues of the developing leaf toughen as the fiddlehead slowly unfurls

The lower fronds begin to photosynthesize as soon as they unfurl

BLECHNUM ORIENTALE

DICKSONIA ANTARCTICA

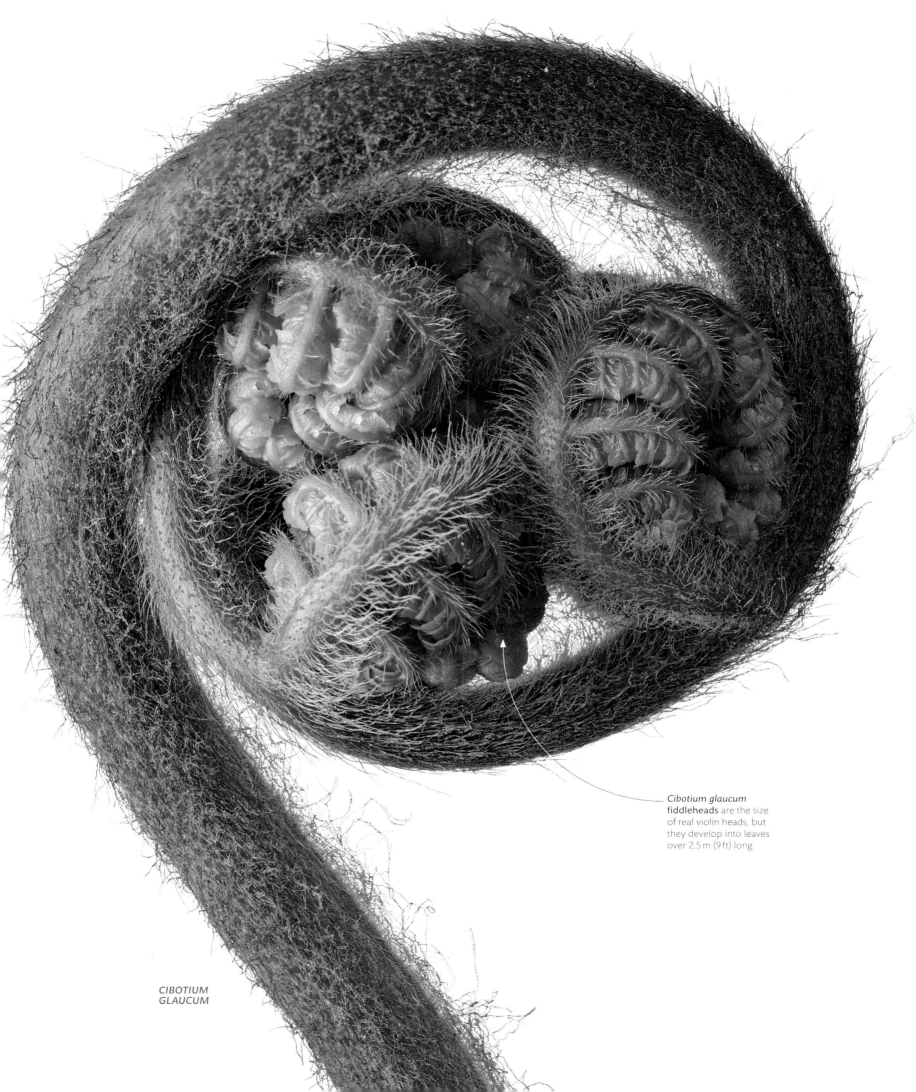

Cibotium glaucum
fiddleheads are the size
of real violin heads, but
they develop into leaves
over 2.5 m (9 ft) long

CIBOTIUM
GLAUCUM

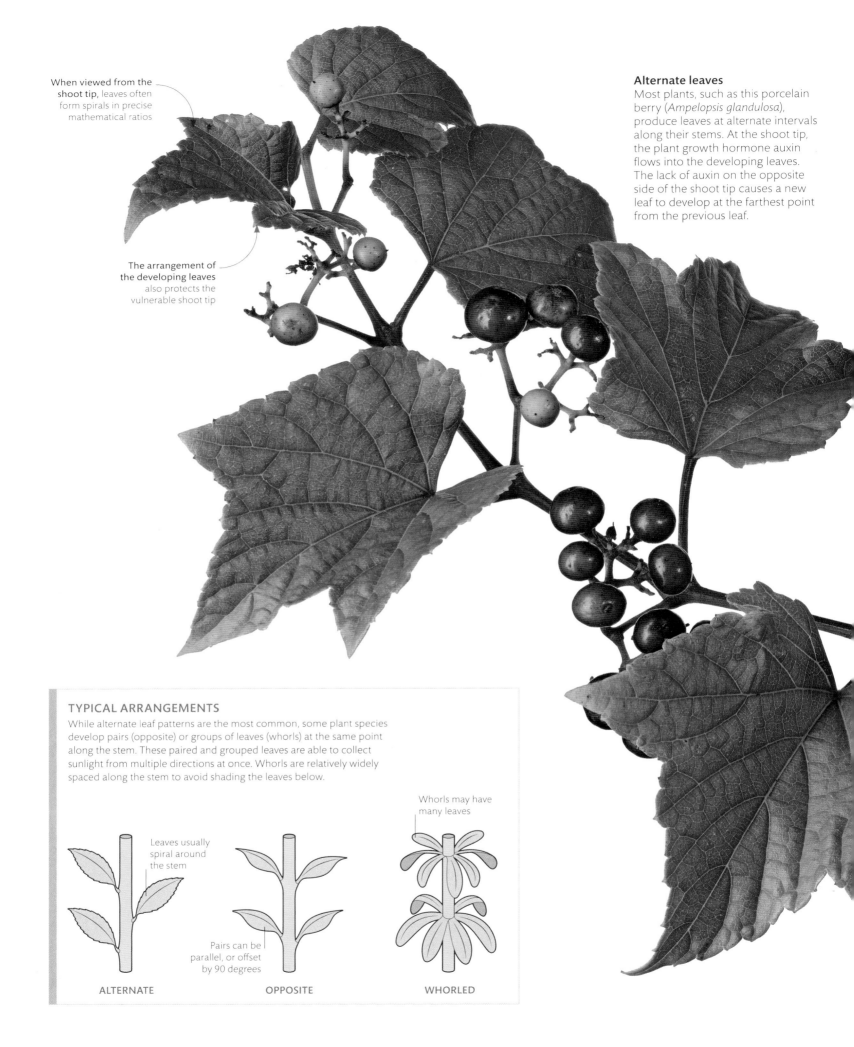

When viewed from the shoot tip, leaves often form spirals in precise mathematical ratios

The arrangement of the developing leaves also protects the vulnerable shoot tip

Alternate leaves

Most plants, such as this porcelain berry (*Ampelopsis glandulosa*), produce leaves at alternate intervals along their stems. At the shoot tip, the plant growth hormone auxin flows into the developing leaves. The lack of auxin on the opposite side of the shoot tip causes a new leaf to develop at the farthest point from the previous leaf.

TYPICAL ARRANGEMENTS

While alternate leaf patterns are the most common, some plant species develop pairs (opposite) or groups of leaves (whorls) at the same point along the stem. These paired and grouped leaves are able to collect sunlight from multiple directions at once. Whorls are relatively widely spaced along the stem to avoid shading the leaves below.

Leaves usually spiral around the stem

Pairs can be parallel, or offset by 90 degrees

Whorls may have many leaves

ALTERNATE

OPPOSITE

WHORLED

leaf arrangement

It is important to avoid being shaded by neighbours, but plants also have to avoid shading their own leaves. Leaf arrangement, or phyllotaxis, occurs in patterns that are specific to each species and prevents the upper leaves from blocking light to the lower branches, enabling the plant to absorb as much sunlight as possible.

The brightly coloured berries are positioned in gaps between the leaves to attract birds and other seed dispersers

Complex composition
Laelia tenebrosa, Philodendron hybrid, *Calathea ornata,*
Philodendron leichtlinii, Polypodiaceae (1989) by Pandora Sellars
not only shows Sellars' exceptional ability to render plants
accurately, but is also an example of her artist's sense of
composition, capturing the way light plays across the leaves.

***Brittany 1979* (detail)**
This watercolour on vellum painting of a fallen
leaf from a European pear tree (*Pyrus communis*),
is a portrait of exquisite beauty. McEwen's focus is
on nature's treasure lying underfoot in this painting from
his series of leaves in autumn colours or stages of decay.
Recording the location and year of each painting, he
captured the hues and imperfections of leaves with
botanic integrity and an artist's sensibility.

plants in art

botanic reinvention

There was a revolution in botanic art when artists stopped seeking perfect
specimens and chose to reveal the beauty in ordinary vegetables, fruit, and
flowers, with all their minute imperfections, and leaves ravaged by beetles
or in stages of decay. The 20th-century British artist Rory McEwen was a
pioneer of this approach and is widely regarded as the first botanical
painter to portray the natural world with the mind of a modern artist.

McEwen's radical reworking of botanical
art emerged from the general ferment of
change in the 1960s. He worked on
vellum, rather than paper. He found that
his watercolours had an extraordinary
translucency and intensity on the vellum's
silky smooth and non-porous surface,
just as illustrations did in medieval
illuminated manuscripts.

McEwen painted with scientific
precision, using tiny, fine brushes, and
applied the same meticulous techniques
to every subject, regardless of whether
it was a heritage bloom, an onion, or a
fallen leaf that he had picked up from the
pavement. He took the time to paint his
subjects in minute detail, highlighting
their beauty of form and colour and
including even so-called imperfections.

Another British artist who took botanical
art to new heights in the 20th century was
Pandora Sellars. Her work appeared in
many botanical publications, and her
artistic ability and sensibility brought
worldwide recognition. Sellars started
her career as a botanical artist when she
found that a camera was incapable of
capturing the colour and form of the
orchids in her husband's greenhouse.

***Indian Onion painted in Benares 1971* (detail)**
This onion, depicted in glowing shades of purple and pink
with a papery brown skin, is almost tangible. Painted in
translucent watercolour, it seems to be poised in space.
McEwen's series of paintings of onions were among his
most influential and fascinating works.

66 A dying leaf should be able to carry the
weight of the world. 99

RORY McEWEN, *LETTERS*

Waxy surfaces and drip tips help rainforest leaves to shed rainwater rapidly

leaves and the water cycle

Plants use less than five per cent of the water they take from the soil; the rest evaporates from the surface of the leaves into the surrounding air. This seemingly wasteful process, known as transpiration, can be a problem in dry climates, but it is vital in many ways. Transpiration makes it possible for water to move up against the pull of gravity through even the tallest trees, carrying with it nutrients from the soil needed for growth. In hot climates, the evaporating water cools the leaves, just as sweating cools human skin.

The red lower leaf surfaces maximize light absorption in shady conditions

TRANSPIRATION

When the stomata are opened to take in carbon dioxide, water continuously evaporates from the leaves. This creates negative pressure, drawing water up the stem from the roots through the plant's vascular system – bundles of fine tubes known as xylem.

Evaporation

Soil

Water

Upper epidermis

Mesophyll cells

Lower epidermis

Stoma

Water

ROUTE OF WATER THROUGH A LEAF

Xylem

Wasting water?

Costus guanaiensis grows in tropical South America. Water is so abundant in the rainforest that plants can produce large leaves to maximize light absorption without the risk of dehydrating due to their high rate of transpiration. As a result, around 30 per cent of all the rain that falls over land each year passes through the leaves of rainforest plants.

Rainforest leaves have many stomata to maximize the uptake of carbon dioxide

The long leaves of *Costus guanaiensis* are typically up to 60 cm (2 ft) in length

leaves and light

A plant's leaves gather sunlight and convert its energy into food by means of a complex process called photosynthesis. Plants use light and a light-sensitive green pigment called chlorophyll in their leaves to transform carbon dioxide from the air and water from the soil into sugars to feed themselves. As a by-product, they also produce the oxygen that supports almost all life on Earth.

The green colour
of leaves comes from chlorophyll, a pigment that absorbs light energy

Broad surface area
Philodendron ornatum leaves are large because the plant grows in the shade and needs to survive with little light. As with most plants, a network of veins transports water absorbed by the roots to the leaf; it also takes the sugars produced by photosynthesis to the rest of the plant.

The long leaf stem
makes it possible for the plant to tilt the leaf towards the sun

Philodendron ornatum **leaves** grow up to 60 cm (2 ft) long, to harvest as much light as possible

TOP OF LEAF

UNDERSIDE OF LEAF

A drip tip makes it easy for rainwater to drain away

The underside of the leaf is dull because it contains fewer chloroplasts

PHOTOSYNTHESIS

Just below the surface of a leaf are specialized cells called mesophyll cells, which carry out photosynthesis. Mesophyll cells contain tiny particles called chloroplasts, which contain the light-absorbing pigment chlorophyll. Chloroplasts harvest light energy from the sun, take carbon dioxide from the air, and water (absorbed from the soil by roots and delivered to the leaf by the plant's vascular system) and convert it all into glucose. This is then packaged into sucrose, a form of sugar that the plant uses as food. During photosynthesis, oxygen is also released into the air through the leaf's pores (stomata).

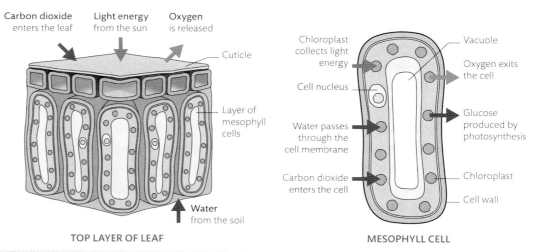

Carbon dioxide enters the leaf

Light energy from the sun

Oxygen is released

Cuticle

Layer of mesophyll cells

Water from the soil

TOP LAYER OF LEAF

Chloroplast collects light energy

Vacuole

Cell nucleus

Oxygen exits the cell

Water passes through the cell membrane

Glucose produced by photosynthesis

Carbon dioxide enters the cell

Chloroplast

Cell wall

MESOPHYLL CELL

Thick tissues and
stiff veins help the
large leaf to maintain
its shape

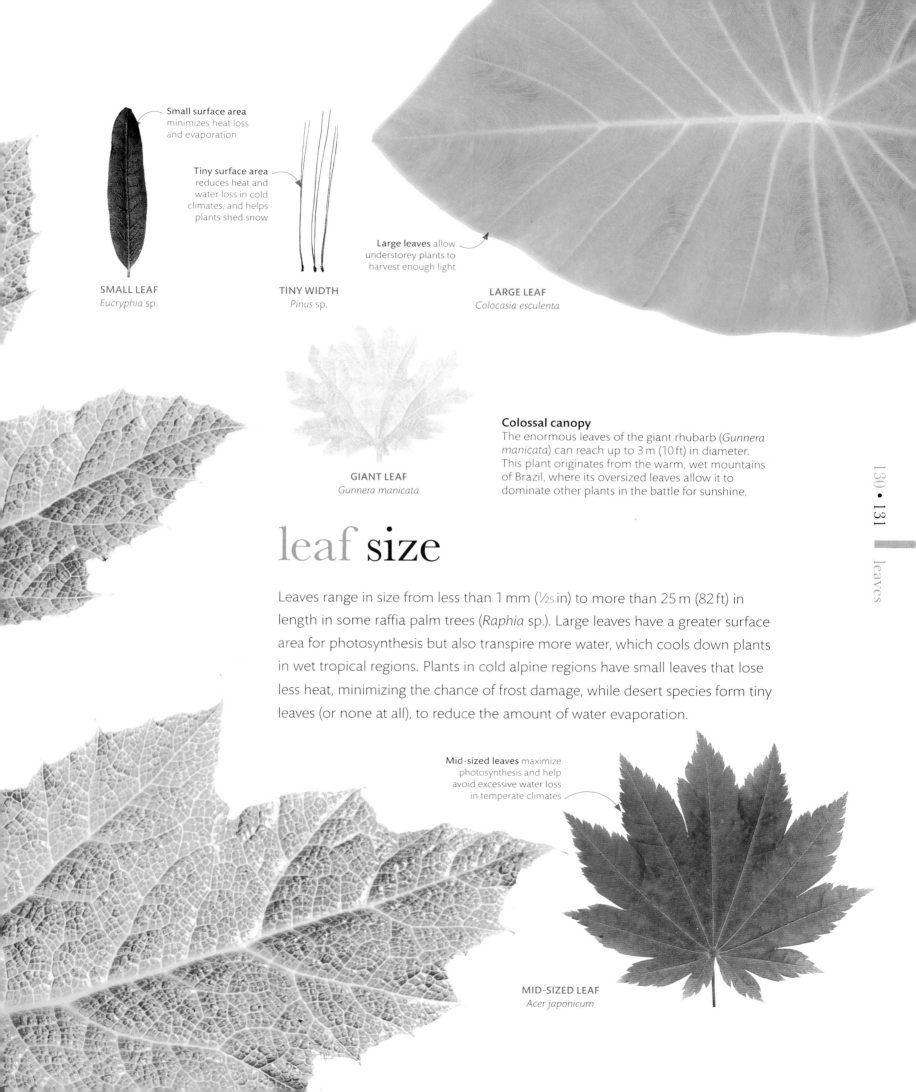

Small surface area minimizes heat loss and evaporation

Tiny surface area reduces heat and water loss in cold climates, and helps plants shed snow

Large leaves allow understorey plants to harvest enough light

SMALL LEAF
Eucryphia sp.

TINY WIDTH
Pinus sp.

LARGE LEAF
Colocasia esculenta

GIANT LEAF
Gunnera manicata

Colossal canopy
The enormous leaves of the giant rhubarb (*Gunnera manicata*) can reach up to 3 m (10 ft) in diameter. This plant originates from the warm, wet mountains of Brazil, where its oversized leaves allow it to dominate other plants in the battle for sunshine.

leaf size

Leaves range in size from less than 1 mm (⅟₂₅ in) to more than 25 m (82 ft) in length in some raffia palm trees (*Raphia* sp.). Large leaves have a greater surface area for photosynthesis but also transpire more water, which cools down plants in wet tropical regions. Plants in cold alpine regions have small leaves that lose less heat, minimizing the chance of frost damage, while desert species form tiny leaves (or none at all), to reduce the amount of water evaporation.

Mid-sized leaves maximize photosynthesis and help avoid excessive water loss in temperate climates

MID-SIZED LEAF
Acer japonicum

leaf shapes

Leaves grow in a wide array of shapes and sizes, each enabling a plant to thrive in its natural habitat. Leaf shapes balance the plant's need to take in light with that of avoiding water loss or resisting damage from wind and rain. Simple leaves grow in one piece, whereas a compound leaf has several parts.

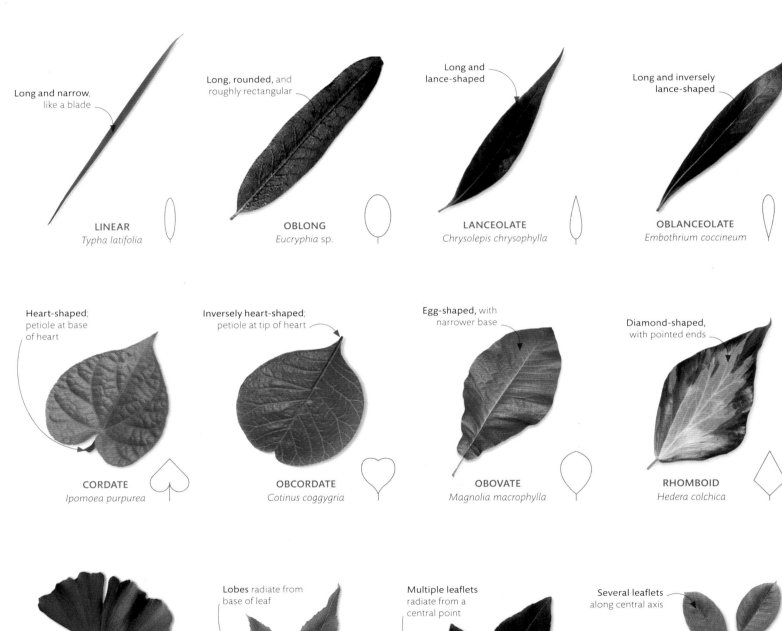

Long and narrow, like a blade

LINEAR
Typha latifolia

Long, rounded, and roughly rectangular

OBLONG
Eucryphia sp.

Long and lance-shaped

LANCEOLATE
Chrysolepis chrysophylla

Long and inversely lance-shaped

OBLANCEOLATE
Embothrium coccineum

Heart-shaped; petiole at base of heart

CORDATE
Ipomoea purpurea

Inversely heart-shaped; petiole at tip of heart

OBCORDATE
Cotinus coggygria

Egg-shaped, with narrower base

OBOVATE
Magnolia macrophylla

Diamond-shaped, with pointed ends

RHOMBOID
Hedera colchica

Divided into two lobes; older leaves fan-shaped

BILOBED
Ginkgo biloba

Lobes radiate from base of leaf

PALMATELY LOBED
Liquidambar styraciflua

Multiple leaflets radiate from a central point

PALMATE COMPOUND
Ptelea trifoliata

Several leaflets along central axis

PINNATE COMPOUND
Juglans regia

Parallel or convergent evolution

Why do species growing in the same environment have similarly shaped leaves? Evolution is a factor. Over time, a plant's DNA changes as individual plants with leaf shapes better suited to their particular needs, and their environment, are the ones that survive and reproduce. Plants with less optimal leaf shapes die. Evolution does not always create a perfect leaf shape, but selects the best option available.

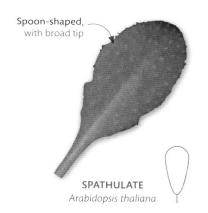

Spoon-shaped, with broad tip

SPATHULATE
Arabidopsis thaliana

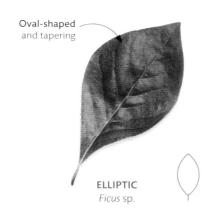

Oval-shaped and tapering

ELLIPTIC
Ficus sp.

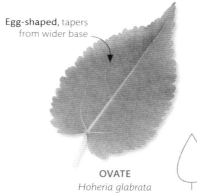

Egg-shaped, tapers from wider base

OVATE
Hoheria glabrata

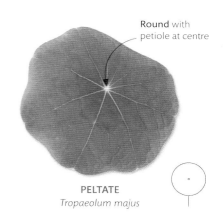

Round with petiole at centre

PELTATE
Tropaeolum majus

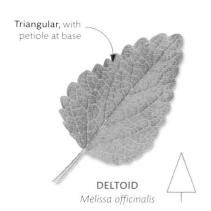

Triangular, with petiole at base

DELTOID
Melissa officinalis

Spear-shaped, with lobes pointing outwards

HASTATE
Hedera sp.

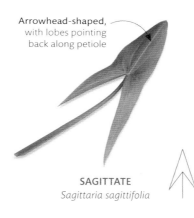

Arrowhead-shaped, with lobes pointing back along petiole

SAGITTATE
Sagittaria sagittifolia

Deeply lobed

PINNATIFID
Quercus sp.

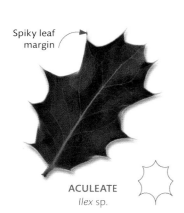

Spiky leaf margin

ACULEATE
Ilex sp.

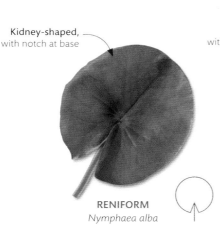

Kidney-shaped, with notch at base

RENIFORM
Nymphaea alba

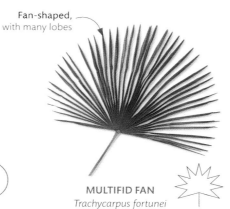

Fan-shaped, with many lobes

MULTIFID FAN
Trachycarpus fortunei

Holes across leaf blade

FENESTRATE
Monstera deliciosa

DESIGNED FOR DRAINAGE

Drip tips are usually aligned with major veins, which channel water from the leaf. Each lobe of more complex leaves, such as those shown on the right or the segments (pinnae) of fern fronds, forms its own drip tip. Combined with a leaf's water-repelling, waxy outer layer (cuticle), drip tips enable rainforest plants to cope with extremely heavy rainfall.

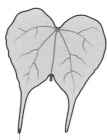

Both lobes have a drip tip

BAUHINIA SCANDENS

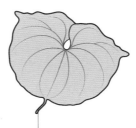

A central drip tip is flanked by two others

DIOSCOREA SANSIBARENSIS

Variable numbers of drip tips per leaf

BEGONIA INVOLUCRATA

Each lobe has a drip tip

MONSTERA DELICIOSA

Queen anthurium

Native to South America, the queen anthurium (*Anthurium warocqueanum*) has leaves that are more than 1 m (3 ft) long. This large surface area receives a lot of rain, but it runs off rapidly with the help of a drip tip. Drip tips are more common in understorey plants such as anthurium than in leaves that grow at the top of the rainforest canopy, which dry rapidly in the sun.

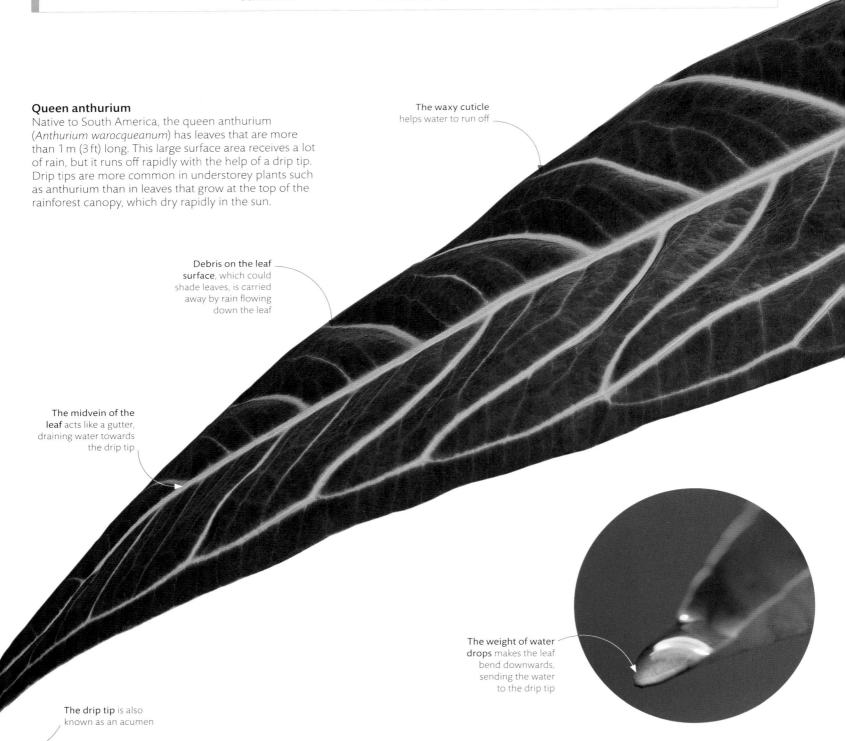

The waxy cuticle helps water to run off

Debris on the leaf surface, which could shade leaves, is carried away by rain flowing down the leaf

The midvein of the leaf acts like a gutter, draining water towards the drip tip

The weight of water drops makes the leaf bend downwards, sending the water to the drip tip

The drip tip is also known as an acumen

The long leaf can turn towards the brightest source of light

A collective vein circles the leaf near its edge, channelling water into the leaf's central gutter. All *Anthurium* species have a collective vein

Leaves on Asian figs have drip tips to cope with heavy rainfall

FICUS SP.

The petiole holds the leaf at an angle, directing water downwards

drip tips

As an adaptation to heavy rainfall, many tropical rainforest leaves form drip tips – elongated leaf tips that allow water to drain off rapidly. The exact benefit of this is unclear. Some researchers think that water left standing on a leaf might encourage the growth of harmful fungi, algae, or bacteria, while others believe that removing the water may help the leaf to regulate its temperature or prevent the droplets from reflecting sunlight, which would hinder photosynthesis.

leaf margins

The edge of a leaf, known as its margin, is a distinguishing feature that can be used to identify plant species. Leaf margin shapes help plants adapt to their environment. Lobed or serrated edges increase the movement of air around the leaf, leading to more water loss but also allowing the leaf to take in more carbon dioxide for photosynthesis. Smooth margins help rainforest plants to shed rainwater rapidly.

Smooth edge, with no serrations (teeth) or indentations

ENTIRE
Eucryphia sp.

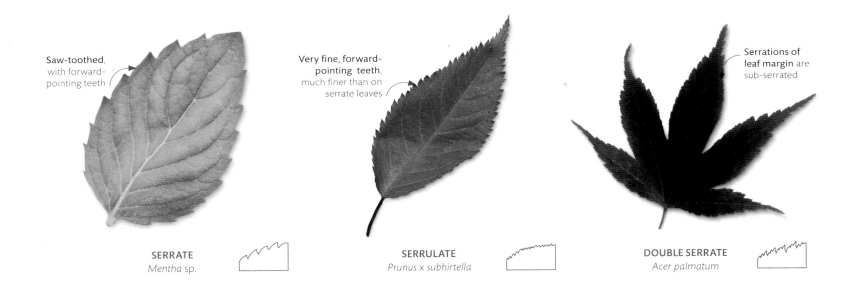

Saw-toothed, with forward-pointing teeth

SERRATE
Mentha sp.

Very fine, forward-pointing teeth, much finer than on serrate leaves

SERRULATE
Prunus x *subhirtella*

Serrations of leaf margin are sub-serrated

DOUBLE SERRATE
Acer palmatum

Edge has scallops, known as crenations

CRENATE
Viola reichenbachiana

Scallops (crenations) are smaller than on crenate leaves

CRENULATE
Cercidiphyllum japonicum

Fine hairs (cilia) cover the margin

CILIATE
Ailanthus altissima leaflet

SINUATE
Quercus macranthera

Undulating (sinuous) edge to leaf

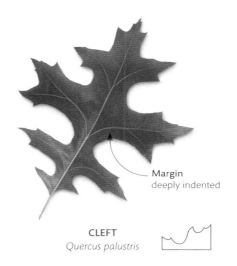

CLEFT
Quercus palustris

Margin deeply indented

Irregular indentations make leaf edge appear torn or cut

INCISED
Koelreuteria paniculata leaflet

Teeth point outwards

DENTATE
Quercus cerris

Very fine, outward-pointing teeth, much finer than on dentate leaves

DENTICULATE
Morus rubra

Rounded indentations on the leaf margin

LOBATE
Quercus petraea

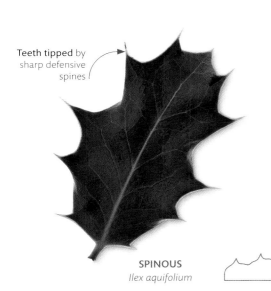

Teeth tipped by sharp defensive spines

SPINOUS
Ilex aquifolium

Margin has 3D waves that make the leaf difficult to flatten

UNDULATE
Philodendron ornatum

Margins and climate

Plants from warm, dry climates often have entire (smooth) leaf margins, which lose less water than leaves with jagged edges. Sap flows faster through serrated leaves, allowing temperate plants to begin photosynthesizing quickly during periods of warm weather. Looking at the margins of fossil leaves can reveal details about Earth's climate at the time when the leaves were alive.

hairy foliage

Deterring herbivores, protecting against extremes of weather, and repelling competing plants with herbicides – plants achieve all this and more using hair-like structures called trichomes on their leaves, stems, and flower buds. Trichomes obstruct insects trying to feed or lay eggs, and may secrete toxins to defend themselves. Some plants with trichomes inject irritating chemicals into the skin of mammals, as a warning to stay away.

STACHYS BYZANTINA

MINT DEFENCES

The trichomes of many plants interfere with insect feeding, but some actively fight these pests. Mint (*Mentha* sp.) trichomes produce the essential oil menthol, which both repels insects and kills those that take a bite.

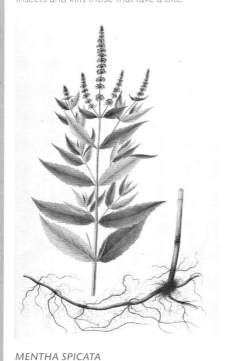

MENTHA SPICATA

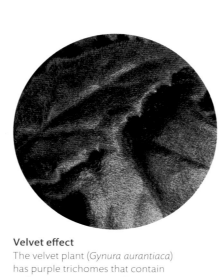

A dense forest of trichomes makes it difficult for insects to move through the mass of hairs to feed on the leaf

Velvet effect
The velvet plant (*Gynura aurantiaca*) has purple trichomes that contain anthocyanin pigments. These pigments prevent the usually shaded leaves from being damaged by flashes of strong sunlight that reach the forest floor.

Defence against the elements
Lamb's ears (*Stachys byzantina*) is covered by a silky layer of hair-like trichomes that help the plant to cope with drought. The hairs trap moisture next to the leaves and deflect the wind, minimizing evaporation, while their silver colour reflects excessive light and heat from the sun.

Fuzzy trichomes are harvested by some bee species, which use them to line their nests

Trichomes cover the leaves, stems, and flower buds, insulating them against frost and heat

Lamb's ears also have glandular trichomes. These excrete compounds with antimicrobial properties to help keep the plant free from disease

Rubus sylvestris s. leninus

plants in art

ancient herbals

Herbals were books or manuscripts containing plant descriptions and information on their properties and medicinal uses. They were also used as reference guides for plant identification and botanical study. They were among the first books and literature ever produced, and some ancient examples contain the earliest known drawings and paintings of plants.

Herbals were probably based on the plant lore and traditional medicines of the ancient world. Some of the earliest examples, from the Middle East and Asia, date back to several thousand years BCE. Herbals became popular in Classical antiquity, the most influential being *De Materia Medica* (*c.*50–70 CE), which was created by Pedanius Dioscorides, a Greek physician in the Roman army. Containing detailed information on more than 500 plants, it was copied extensively and used continuously for more than 1,500 years. It is not known whether the original version was illustrated, but the *Vienna Dioscorides*, considered to be the oldest manuscript version, features naturalistic and often minutely detailed paintings.

Woodblock and woodcut printing increased the scope for reproduction, but it was the invention of the printing press in the 15th century that led to a profusion of illustrated herbals and an improvement in the quality of the images. Although their popularity eventually diminished, herbals may be seen as forerunners of the scientific books with botanically accurate illustrations that replaced them.

Culpeper's herbal
This handcoloured botanical copperplate engraving features in *The English Physitian* by Nicholas Culpeper (1652). Affordable, accessible, and practical, it became one of the most popular and successful books of its type.

De Materia Medica
This page from Dioscorides' *De Materia Medica* illustrates a wild bramble (with the Greek word *batos* just above) identified as *Rubus sÿlvestris*. This copy of the manuscript was produced in 1460, nearly 1,400 years after the original was first produced. It was part of the collection owned by renowned British botanist and naturalist Sir Joseph Banks.

       ... one of the rare types of manuscript with an almost continuous line of descent from the time of the ancient Greeks to the end of the Middle Ages. ,,

MINTA COLLINS, *MEDIEVAL HERBALS: THE ILLUSTRATIVE TRADITIONS*, 2000

Jade necklace

The tightly stacked leaves of the jade necklace plant (*Crassula rupestris* subsp. *marnieriana*) are small and rounded to reduce their surface area and minimize the evaporation of water, which they store within specialized cells. Their cuticles are often coated with a white waxy "bloom", which reflects damaging heat and light from the sun.

The tightly packed leaves resemble groups of lumpy stems

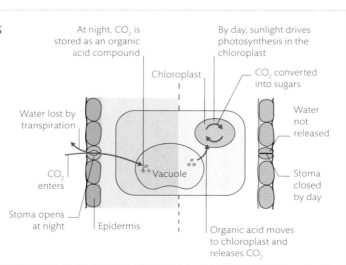

succulent foliage

Few plants can survive for long without water, but succulent plants store water in their thickened leaves or stems and are specially adapted to conserve every drop. Not only do the leaves have a dense, waterproof, waxy cuticle, but their stomata (tiny pores) are often sunken, to decrease the airflow and increase the humidity around them. Unlike most plants, succulents open their stomata at night, to minimize loss of water by evaporation during the heat of the day.

CAM PHOTOSYNTHESIS

To conserve water, succulents use a type of photosynthesis called crassulacean acid metabolism (CAM). Instead of taking in carbon dioxide (CO_2) during the day, CAM plants open their stomata at night to reduce transpiration (see p.126). The carbon dioxide is stored as an organic acid compound, then moved to the chloroplast during the day and released for use in photosynthesis.

At night, CO_2 is stored as an organic acid compound

By day, sunlight drives photosynthesis in the chloroplast

Chloroplast

CO_2 converted into sugars

Water lost by transpiration

Water not released

CO_2 enters

Vacuole

Stoma closed by day

Stoma opens at night

Epidermis

Organic acid moves to chloroplast and releases CO_2

INSIDE A MESOPHYLL CELL

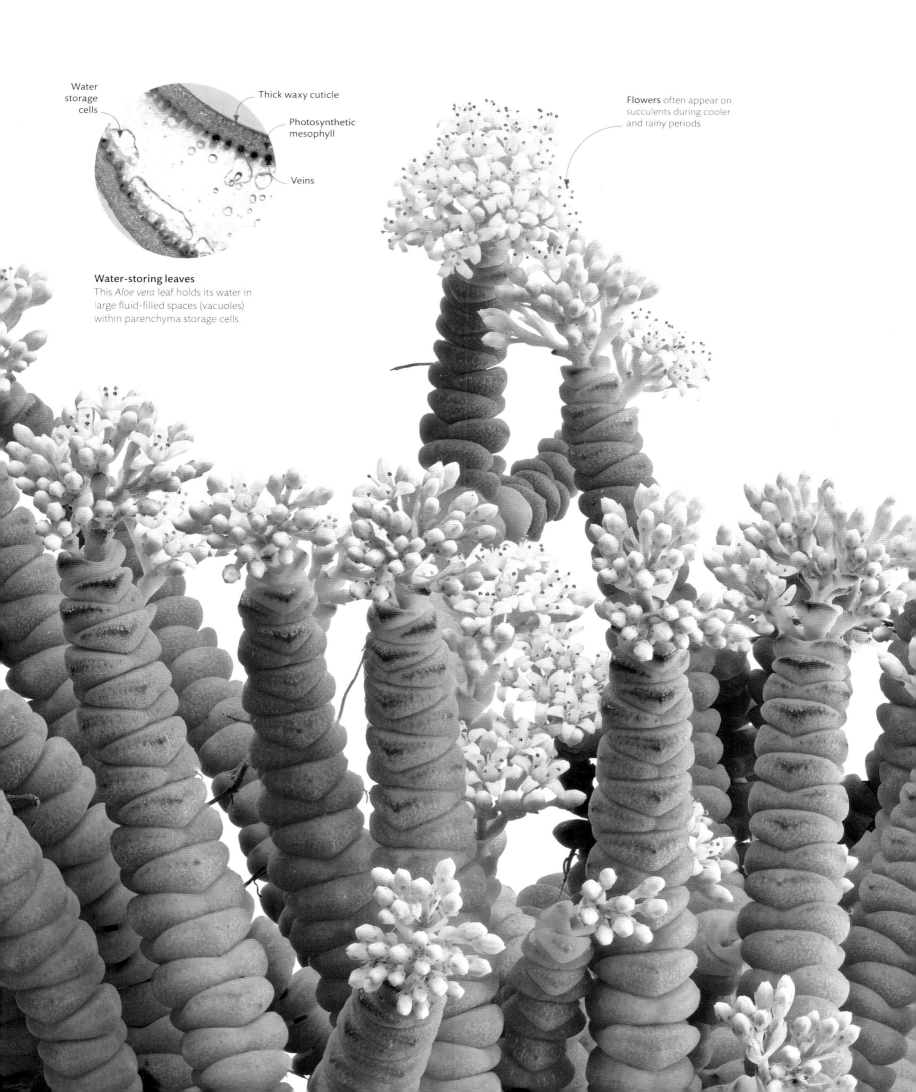

Water
storage
cells

Thick waxy cuticle

Photosynthetic
mesophyll

Veins

Water-storing leaves
This *Aloe vera* leaf holds its water in
large fluid-filled spaces (vacuoles)
within parenchyma storage cells.

Flowers often appear on
succulents during cooler
and rainy periods

Self-cleaning leaves

The leaves of the sacred lotus (*Nelumbo nucifera*) are covered in microscopic bumps and a water-resistant waxy cuticle. Water droplets are repelled by this protective layer and quickly shed, but they pick up dirt particles as they run along the leaf, cleaning its surface and ensuring that light can reach the photosynthetic cells beneath. This useful property has been replicated in laboratories to create self-cleaning coatings.

waxy leaves

Plants first evolved in water, but around 450 million years ago they began to colonize the land. To prevent dehydration, they developed a waterproof, waxy coating called a cuticle over their leaves and stems. The cuticle also protects a plant from infection by microorganisms. Although it is translucent to let light in for photosynthesis, it also reflects excessive light and heat, to prevent damage to the plant.

IN CLOSE-UP

Waxy cuticles are made of water-resistant chemical compounds. These help to prevent water from evaporating from the leaf while also providing some protection against fungi and bacteria. The fur-like covering on the euphorbia plant below is wax crystals, which often form on the outer surface of the cuticle.

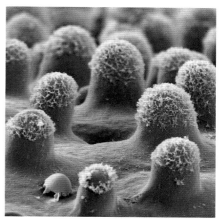

EUPHORBIA LEAF SURFACE

The umbrella-like lotus leaves measure up to 60 cm (2 ft) across

The surface of a sacred lotus leaf is self-cleaning, keeping it free of dirt even in muddy habitats

A long petiole connects the leaves to roots at the bottom of the pond

Banishing dew

Unlike lily pads, lotus leaves often stand above the water on their slender petioles. As the large, balancing leaves vibrate, they cause the dew that forms between the water-repelling bumps on the leaf surface to run off.

Aloidendron dichotomum

the quiver tree

The name quiver tree was coined by the San people of southern Africa, who hollowed out this plant's branches and used them as quivers for their arrows. A large succulent, *Aloidendron dichotomum* is a very hardy species that can grow to 7 m (23 ft) in height and live for more than 80 years.

Native to southern Namibia and the Northern Cape region of South Africa, the quiver tree is essentially an aloe that grows as big as a tree. Like its smaller relatives, it produces the rosettes characteristic of an aloe, but at the tips of its forking branches.

The quiver tree's branches are covered in a powdery white substance that acts like a form of protective sunscreen. As temperatures rise and the surrounding landscape sizzles beneath the fierce South African sun, the powder helps the quiver tree to maintain its internal temperature at a much more tolerable level.

In spring, long spikes with bright orange-yellow flowers emerge from each rosette of leaves. The spikes are like flags signalling to wildlife from far and wide. From bees to birds and even baboons, many animals arrive to feast on the quiver tree's nectar. Even when it is not in flower,

the tree provides valuable nesting sites for birds among its sturdy branches. Mature quiver trees often boast huge colonies of sociable weaver birds, which weave large, intricate communal nests among the branches, taking advantage of what little shade they can provide.

Prolonged droughts have led to many quiver trees dying in the hotter parts of their native range. With climate change, such droughts are predicted to become more widespread and severe. The loss of the quiver tree would be an indication of serious changes in rainfall.

Forking habit
Although it grows into a tree, this plant does not produce wood. Instead, its stout branches are filled with pulpy fibres that store precious water. As a quiver tree grows, its branches divide, producing two new branches each time.

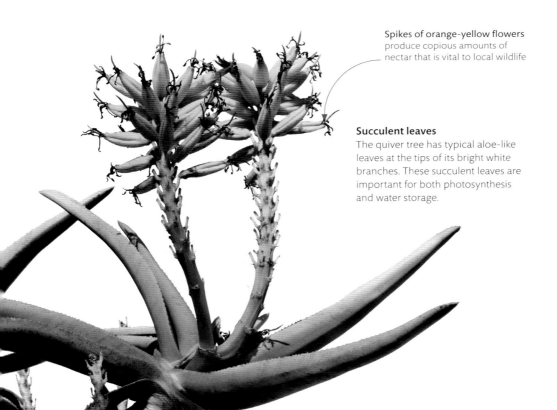

Spikes of orange-yellow flowers
produce copious amounts of nectar that is vital to local wildlife

Succulent leaves
The quiver tree has typical aloe-like leaves at the tips of its bright white branches. These succulent leaves are important for both photosynthesis and water storage.

silver leaves

In dry mountain climates with intense sunlight and high evaporation, many plants keep cool by having silvery leaves, which deflect light and heat. The colour comes from a coating of translucent wax or hairs (trichomes) on top of the green leaf cells underneath. Both the wax and the hairs also help to reduce water loss: hairs raise the humidity around the leaf's surface, which minimizes evaporation, and the wax creates an additional layer of waterproofing.

EFFICIENT GAS EXCHANGE
The Tasmanian snow gum (*Eucalyptus coccifera*) has leaves that grow vertically, to minimize their exposure to the sun's heat. It also means that each leaf can have stomata (pores) on both sides and take in more carbon dioxide for photosynthesis, without losing too much water from evaporation.

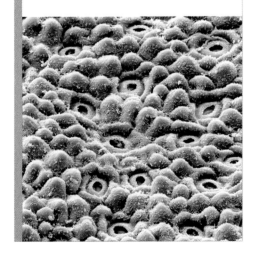

Wax is produced as soon as the leaf begins to form

The dense waxy covering is difficult for insect pests to walk on

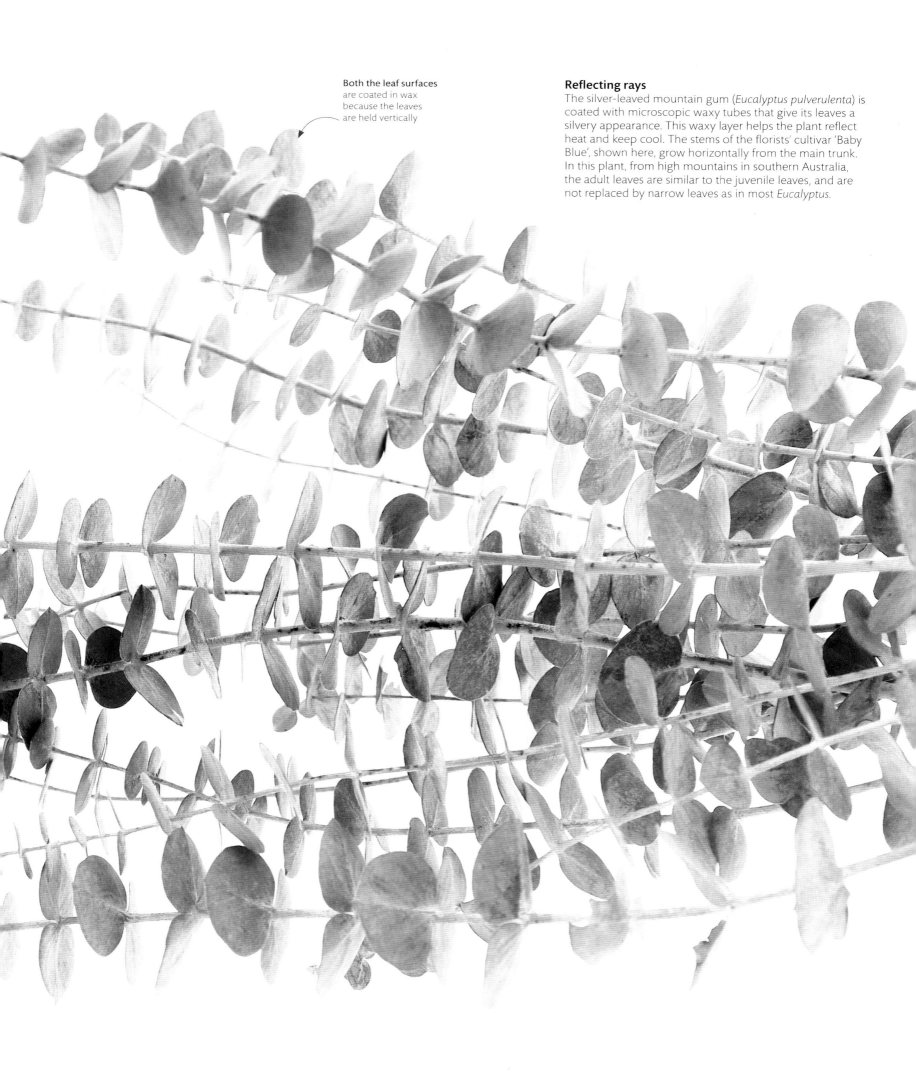

Both the leaf surfaces are coated in wax because the leaves are held vertically

Reflecting rays

The silver-leaved mountain gum (*Eucalyptus pulverulenta*) is coated with microscopic waxy tubes that give its leaves a silvery appearance. This waxy layer helps the plant reflect heat and keep cool. The stems of the florists' cultivar 'Baby Blue', shown here, grow horizontally from the main trunk. In this plant, from high mountains in southern Australia, the adult leaves are similar to the juvenile leaves, and are not replaced by narrow leaves as in most *Eucalyptus*.

variegated leaves

Leaves that are two or more colours are known as variegated. Although common in gardens, these leaves are rare in nature because only the green parts of a leaf can photosynthesize. Some rainforest plants became variegated to avoid damage from sunlight flashing through the trees, or to mimic the effects of disease in order to deter plant-eaters. Most variegated garden plants are chimeras – the different coloured parts of their leaves contain cells that are genetically different.

Light-reflecting air spaces between the outer "skin" and the cells beneath

This spirally variegated begonia was developed for gardeners

Fittonia is known as the nerve plant due to its variegated veins

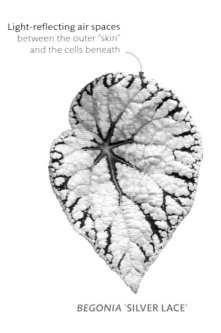

BEGONIA 'SILVER LACE'

BEGONIA 'ESCARGOT'

FITTONIA SP.

Pale patches indicate cells that cannot photosynthesize

Cream and green areas contain cells that are genetically different

Different colours show that different leaf pigments are present

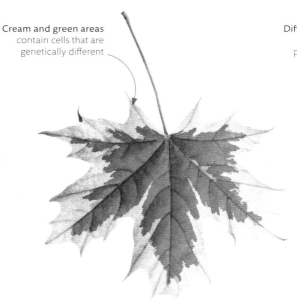

CALATHEA BELLA

ACER PLATANOIDES 'DRUMMONDII'

CODIAEUM VARIEGATUM

Mimicry

Angel wings (*Caladium bicolor*) is native to the forests of Central and South America. The white and red patches on its leaves cannot photosynthesize, but they make the leaf look as if it has been infested by leaf miner insects. This form of deception seems to be very effective: in a related species, variegated leaves like this were found to be up to 12 times less likely to be infested by leaf miners.

White marks mimic the tunnels of leaf miner larvae

CALADIUM BICOLOR

Variegated red leaves need more light than purely green leaves, as they photosynthesize less efficiently

Fake leaf-miner tunnels deter plant-eaters from attacking what appears to be an infested leaf, and inferior food source

AGLAONEMA SP.

CALADIUM BICOLOR

Bred to be red
Native Japanese maples have green leaves, but some cultivars display red leaves all summer. The leaves turn an even brighter red in autumn.

Palmate (hand-shaped) leaves have 5, 7, or 9 serrated and pointed lobes

Acer palmatum

the japanese maple

Few leaves are as recognizable as those of the Japanese maple (*Acer palmatum*). Native to Japan, North and South Korea, and Russia, this attractive deciduous tree has become a mainstay in gardens worldwide, thanks to its graceful shape, elegant leaves, and colourful foliage.

The Japanese maple is a relatively small, slow-growing tree that rarely exceeds 10 m (33 ft) in height, which means it is quite at home growing in the shade of taller neighbours. In the wild, it grows in the understorey of temperate woodlands and forests up to 1,100 m (3,600 ft).

One of the most intriguing features of this tree is how variable its appearance can be. Different native populations take on a range of forms, from diminutive shrubs to spindly trees. Similarly, the leaves that have made it so famous are also quite variable in shape and colour. All of this has to do with the species'

Small wonder
Some Japanese maples have upright forms; others are weeping and often develop a domed-shaped habit. In autumn, the leaves turn startling shades of yellow, orange, red, purple, and bronze before they are shed.

DNA. Japanese maples show a fairly high degree of genetic diversity, and seeds from the same plant can produce offspring that look wildly different from one another.

Plant breeders have taken advantage of this natural variety, creating over 1,000 different cultivars since the 18th century. Much of that breeding has focused on producing bright red foliage. Such vivid coloration comes from a group of plant pigments called anthocyanins. The pigments help to protect leaf tissues from exposure to too much UV radiation or harmful swings in temperature. There is some evidence that anthocyanins may also act as a deterrent to insect herbivores.

Japanese maples produce small, red or purple flowers that are pollinated by wind or insects. These are followed by winged seeds called samaras (see p.314). Like little helicopters, the seeds catch the breeze and twirl off and away from their parents.

autumn colours

As summer draws to a close, the shorter days and cooler temperatures alert trees and shrubs that they need to prepare for winter. Spectacular displays of colour may follow when chemical changes within the leaves gradually transform them from green to brilliant shades of yellow, orange, and red.

Colour combinations
Leaves are coloured by chemical compounds. The most visible of these is chlorophyll, which makes leaves appear green. Carotenoids are responsible for yellow and orange hues, while anthocyanins create shades of red and purple.

Yellow and orange shades are caused by carotenoids, which degrade more slowly than green chlorophyll

Chlorophyll absorbs the red and blue-violet parts of sunlight but reflects green light, giving leaves their normal colour

Green chlorophyll degrades in bright light, but it is constantly replenished by new chlorophyll during the growing season

Chloroplasts are where photosynthesis takes place, using light energy captured by chlorophyll

Yellow carotenoids are always present, but they become more visible when chlorophyll is broken down in autumn leaves

Inside a leaf cell
Leaf cells contain several pigments, but green chlorophyll dominates during spring and summer, when photosynthesis is at its height. The chlorophyll is found in special cell structures called chloroplasts, which produce the plant's food.

Common carotenoids, such as beta-carotene, look orange because they absorb blue and green light and reflect red and yellow

HOW COLOURS CHANGE

In autumn, the plant produces hormones that tell its leaf tissues to die. The chlorophyll in the leaves breaks down more quickly than the carotenoids, so the leaves turn yellow. Meanwhile, red-purple anthocyanins are actively produced in large cell compartments called vacuoles. They give the leaf an orange hue, which turns to red as the last of the carotenoids are lost.

Green leaf turns yellow, then red

Cuticle

Anthocyanins in vacuole

Upper epidermis

Palisade cell

Air space

Lower epidermis

Stoma

Vascular bundle (vein) delivers hormones

Red colours occur when the tree sap is acidic, purple shades if the sap is alkaline

Red-purple colours occur when leaf starches break down into sugars, which react with other chemicals to form anthocyanins

Anthocyanins may act as a natural sunscreen that protects leaves until their nutrients have been recycled for use in other parts of the plant

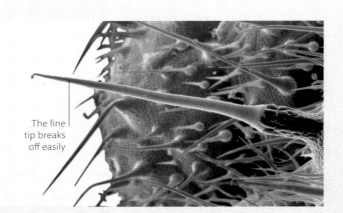

CHEMICAL COCKTAIL
Stinging nettle trichomes contain an arsenal of irritating chemicals, including formic acid, histamine, serotonin, and acetylcholine. These compounds work together to make the pain and discomfort that they cause last much longer than each chemical could achieve individually.

The fine tip breaks off easily

stinging leaves

Plants that sting, such as nettles, are covered in single-celled hairs (trichomes) full of chemicals that cause pain, inflammation, and irritation in animals unlucky enough to brush against them. The brittle trichome tip breaks off easily when touched, revealing a sharp hypodermic needle that immediately injects its toxic mixture under the skin of browsing herbivores. If nettle leaves are damaged by herbivores, they respond by producing more trichomes.

Nettle stems and stalks also carry stinging trichomes

Many more stinging trichomes are located along the veins on the underside of the leaf

Short, non-stinging trichomes provide defensive barriers against insects

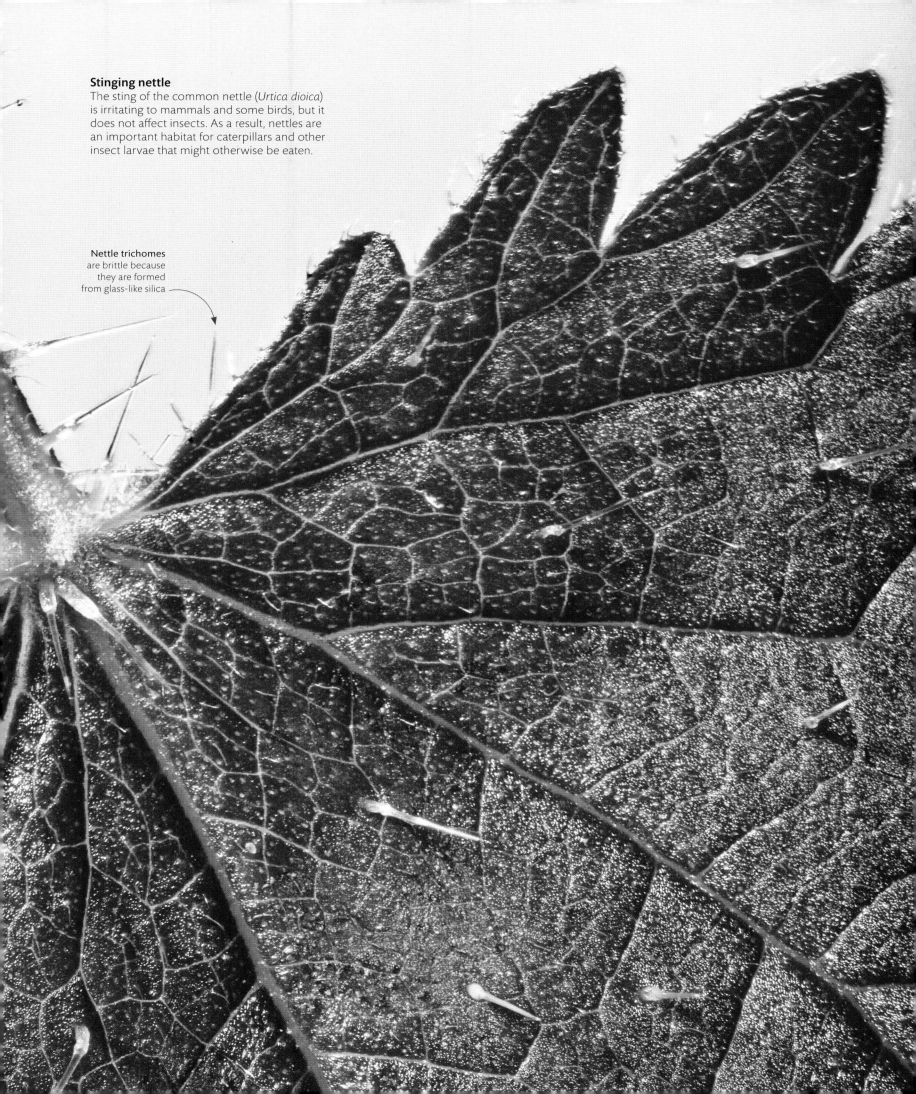

Stinging nettle
The sting of the common nettle (*Urtica dioica*) is irritating to mammals and some birds, but it does not affect insects. As a result, nettles are an important habitat for caterpillars and other insect larvae that might otherwise be eaten.

Nettle trichomes are brittle because they are formed from glass-like silica

Vachellia karroo

the sweet thorn

Formerly called *Acacia karroo*, this fragrant acacia is one of the hardiest trees in southern Africa. A highly adaptable species, it is equally at home in humid forest, on savanna, or in semidesert. Once established, the sweet thorn can handle almost any adversity – even wildfires.

The tree's daunting "thorns" – technically spines, developed from leafy outgrowths (stipules) at the base of leaf-stalks – are up to 5 cm (2 in) long. Being dressed in spines is enough to put off many, but not all, grazing animals. Giraffes, in particular, have no problem wrapping their leathery tongues around its branches to feed on the fine, mimosa-like leaves. The bark, flowers, and nutritious seedpods also provide sustenance to animals, as does the gum that oozes from wounds in the bark, a special favourite of vervet monkeys and lesser bushbabies.

By no means a large tree – bigger specimens reach around 12 m (40 ft) in height – the sweet thorn is also relatively short-lived, with a maximum lifespan of 30 to 40 years. It is, however, able to cope with extreme conditions. As well as being frost-resistant, the sweet thorn can survive drought thanks to its long taproot, which enables it to draw on water reserves deep underground. When nutrients are scarce, it can generate its own supply of nutrients by using structures on its roots that house nitrogen-fixing bacteria (see p.33).

Fast-growing and tolerant of many soil types, the sweet thorn is able to establish itself without shade or shelter, and is even impervious to fire. Seedlings that have survived their first year can be burnt to a crisp only to sprout new stems, thanks to the energy stored in their roots.

The sweet thorn's adaptability makes it an aggressive, invasive species when it is introduced outside its native range. The fact that few animals are brave enough to graze its well-protected foliage also helps it outcompete other vegetation.

Wall of thorns
Sweet thorns are leafless during the winter, when the density of their long, white thorns is clearly visible. These spines make it a favourite nesting spot for birds, as they deter all but the most persistent nest predators.

Each pompom inflorescence is made up of many individual flowers

Pompom display
In early summer, the sweet thorn's canopy erupts with hundreds of yellow pompom-shaped blooms. The tree's long flowering season offers a reliable source of pollen and nectar to bees, making this an important species for honey production.

STAYING SAFE WITH SPINES
Acacia trees often grow large spines that provide a home to beneficial ants, which viciously protect their host from attack. Most cacti have converted all of their leaves into spines; they photosynthesize using their succulent stems.

ACACIA (*ACACIA SPHAEROCEPHALA*)

CACTUS (*MAMMILLARIA INFERNILLENSIS*)

leaf defences

Running away is not an option, so plants have evolved many ways to deter would-be predators. Some have converted their leaves into sharp spines, which hurt animals that attempt to take a bite. Spines can be modified parts of the leaf, such as the vascular tissue, petiole, or stipules, while other plants have converted the entire leaf into defensive spines.

The flowers are unusual for a thistle, in that they resemble daisies when open

The leaf undersides are cobwebby, covered with soft, white hairs

Main stem leaves form broad, spiny "wings"

Upper leaf surfaces are almost shiny

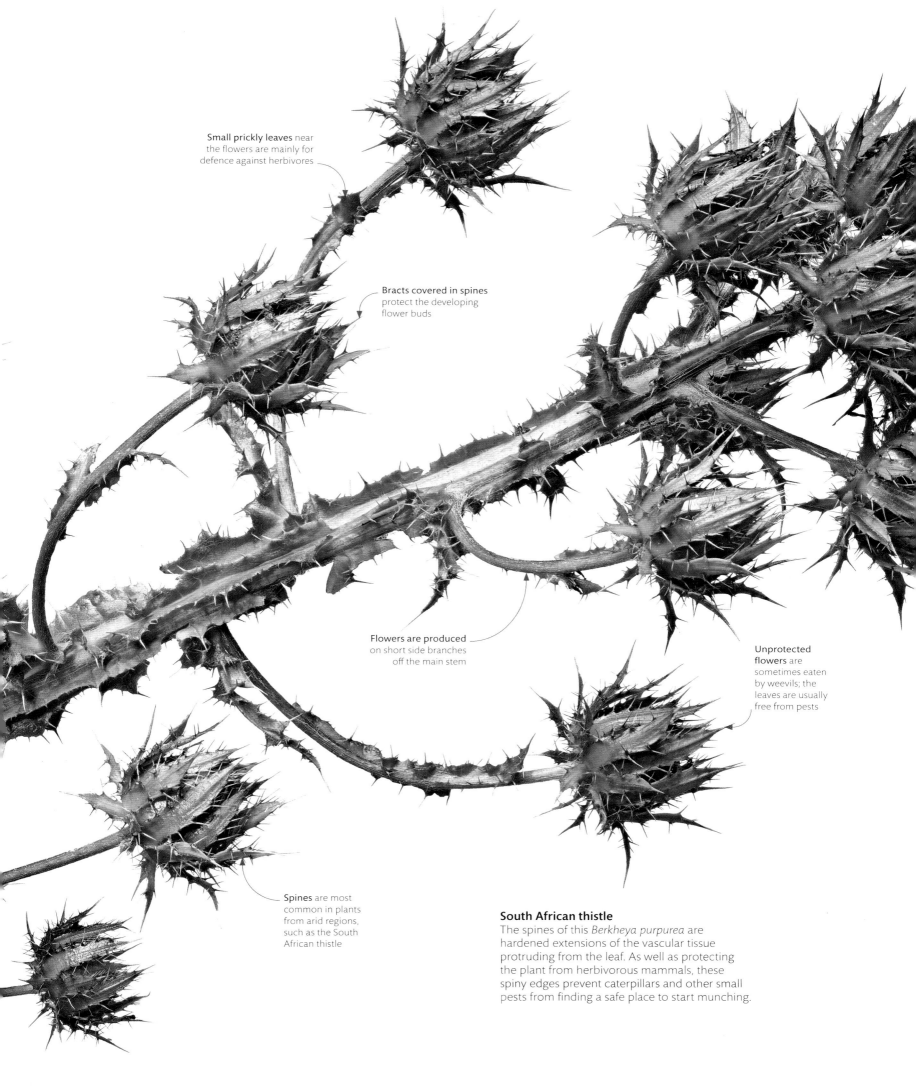

Small prickly leaves near the flowers are mainly for defence against herbivores

Bracts covered in spines protect the developing flower buds

Flowers are produced on short side branches off the main stem

Unprotected flowers are sometimes eaten by weevils; the leaves are usually free from pests

Spines are most common in plants from arid regions, such as the South African thistle

South African thistle
The spines of this *Berkheya purpurea* are hardened extensions of the vascular tissue protruding from the leaf. As well as protecting the plant from herbivorous mammals, these spiny edges prevent caterpillars and other small pests from finding a safe place to start munching.

Surviving the storm

The coconut palm (*Cocos nucifera*) is extremely wind tolerant, its pinnate leaves allowing strong winds to pass right through them. In violent storms, fronds facing the wind may snap off at the base, but this does not usually damage the rest of the plant.

Fronds can reach 5.5 m (18 ft) in length

Leaves surround and protect the delicate growing tip at the top of the stem

A pinnate leaf has leaflets arranged in pairs along a central axis

leaves in extremes

During a hurricane, the leaves of many trees catch the wind like sails, and branches or trunks may snap. Palm trees, however, can be bent horizontally by extreme winds and still emerge largely unscathed. They survive storms thanks to their flexible stems and aerodynamic leaves. Most palms have feather-like pinnate leaves, with a strong, bendy midrib and leaflets that fold up in the wind to avoid major damage.

The leaflets of large fronds fold up like a fan to reduce the area exposed to gales

The trunk is flexible thanks to its vascular tissues that allow it to bend without breaking

the bismarck palm

Offering welcome shade under its impressive canopy, the Bismarck palm is perhaps the most elegant and spectacular of all the fan palms. *Bismarckia nobilis* hails from the arid grasslands of northwestern Madagascar. It is one of the few endemic Madagascan species not currently in decline.

Named after the first chancellor of the German Empire, Otto von Bismarck (1815–98), this tree is the sole member of its genus. *Bismarckia nobilis* is not the tallest of palms, but given time larger individuals can up grow to 18 m (60 ft), though it can take a century to reach such heights.

In its native land, the Bismarck palm must endure extremes of climate, so it has to be hardy. The dry season heralds punishing heat, relentless sun, little to no rain, and wildfires; the wet season brings lots of rain and high humidity. When the weather is driest and the sun fiercest, the tree uses its deep root system to access water far below ground, while the waxy coating that gives its leaves their silvery hues acts like sunscreen and protects the sensitive photosynthetic structures inside from an overdose of solar radiation. A stout trunk combined with leaves and growth tips positioned far above the reach of most flames enables the Bismarck palm to survive all but the very worst wildfires. To make the most of the rain when it does come, the curved frond petioles channel water down to the base of the trunk.

This palm is dioecious, meaning that individual trees produce either male or female flowers, never both. Pollination is achieved by insects and, if male and female trees are growing close enough together, via wind as well. Only the females trees will grow fruit, with each small flower developing a single fleshy, but inedible, stone fruit.

Massive fronds

Fan palms are so named for the shape of their fan-shaped leaf fronds. The Bismarck palm's rounded, silvery blue to green fronds can reach up to 3 m (10 ft) across. Each frond is made up of a spray of multiple stiff, sharp-edged leaflets.

Small, creamy white flowers form on long stalks on a female inflorescence and eventually droop down under the weight of the resulting fruit

***Bismarckia* inflorescence**
The Bismarck palm develops long, rope-like inflorescences made up exclusively of either male or female flowers. Large clusters of fruits form on the female inflorescences.

The leaf is broad so that the upper
surface can collect the maximum
amount of sunlight

Firm anchors

Although the leaves of giant water lilies (*Victoria
amazonica*) float on the water surface, they are
firmly anchored to underground stems at the
bottom of the pond. The formidable prickles
along the stalk and leaf surface protect the
leaves from being eaten by fish.

Floating leaves provide
shelter for young fish,
insects, and amphibians,
but as they have little
competition the leaves
can soon cover a large
area of water, harming
other aquatic life by
blocking light and oxygen

floating leaves

While most plants fight for space on land, some have turned to
a life afloat. Aquatic plants, such as the giant water lily, benefit
from unlimited access to water and less competition for light and
nutrients. Their floating leaves use pockets of air as buoyancy aids,
trapped either in air chambers within the leaf or between dense
hairs on their surface. The leaves of the giant water lily are so
buoyant that they can support the weight of a human baby.

Purple tissues on the lower surface of the leaf keep it warm by absorbing excess sunlight passing through the photosynthetic cells above

Prominent veins

The long leaf stalk connects to a stem buried in the mud

Leaf underside

Water lilies lose a lot of water through evaporation. Taking up more water means absorbing dissolved toxic compounds such as heavy metals, which are safely stored in epidermal glands. Prominent veins are surrounded by thick-walled supporting cells that stiffen the structure of these broad leaves.

HOW WATER LILIES FLOAT

Large air chambers inside water lily leaves provide them with the buoyancy they need for a floating life. Rigid, star-shaped structures called sclereids within the spongy leaf tissues help to maintain the shape of the leaves. This, in turn, enables the leaves to use the water's surface tension to stay afloat.

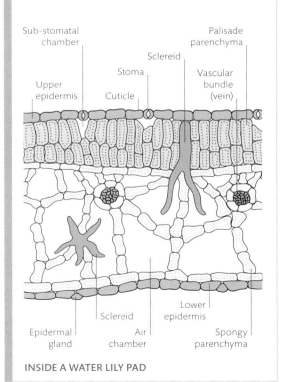

Sub-stomatal chamber

Palisade parenchyma

Sclereid

Stoma

Vascular bundle (vein)

Upper epidermis

Cuticle

Epidermal gland

Sclereid

Air chamber

Lower epidermis

Spongy parenchyma

INSIDE A WATER LILY PAD

A waxy surface helps water to drain easily into the central cavity

plants with pools

Tadpoles living in trees might sound bizarre, but some tree frogs lay their eggs in pools of water held between the leaves of plants such as this bromeliad, *Neoregelia cruenta*, which grows on trees in the rainforest. These tiny pools, or phytotelmata, can also support insects, nematodes, and even small crabs. Phytotelmata can be found in crevices and cavities between leaves, in *Nepenthes* and *Sarracenia* pitcher plant traps, tree hollows, and within bamboo stems.

Wide, concave leaves funnel rainwater towards the central cup, ensuring the plant has a steady water supply

leaves that eat

Carnivorous plants capture and consume prey to obtain nutrients lacking in the soil. They have developed ingenious ways of ensnaring their victims. Venus fly traps snap shut to imprison insects, aquatic bladderworts create a partial vacuum to suck in swimming creatures, and the sticky hairs of sundews fold around prey, holding it tight as it is digested. Insects drown in the liquid at the bottom of pitcher plant cups, while corkscrew plants guide their unsuspecting quarry into a digestion chamber.

Pitcher plants
There are two main pitcher plant families, the Sarraceniaceae and the Nepenthaceae (shown here). Insects that fall into a pitcher plant cup are slowly dissolved. Phosphorus and nitrogen are the main nutrients that pitcher plants gain from their digested prey. Instead of being eaten, the larvae of some insects are adapted to live in the water inside pitchers.

The outer rim, or lip, of the pitcher is slippery when moistened by rain or nectar

NEPENTHES VEITCHII

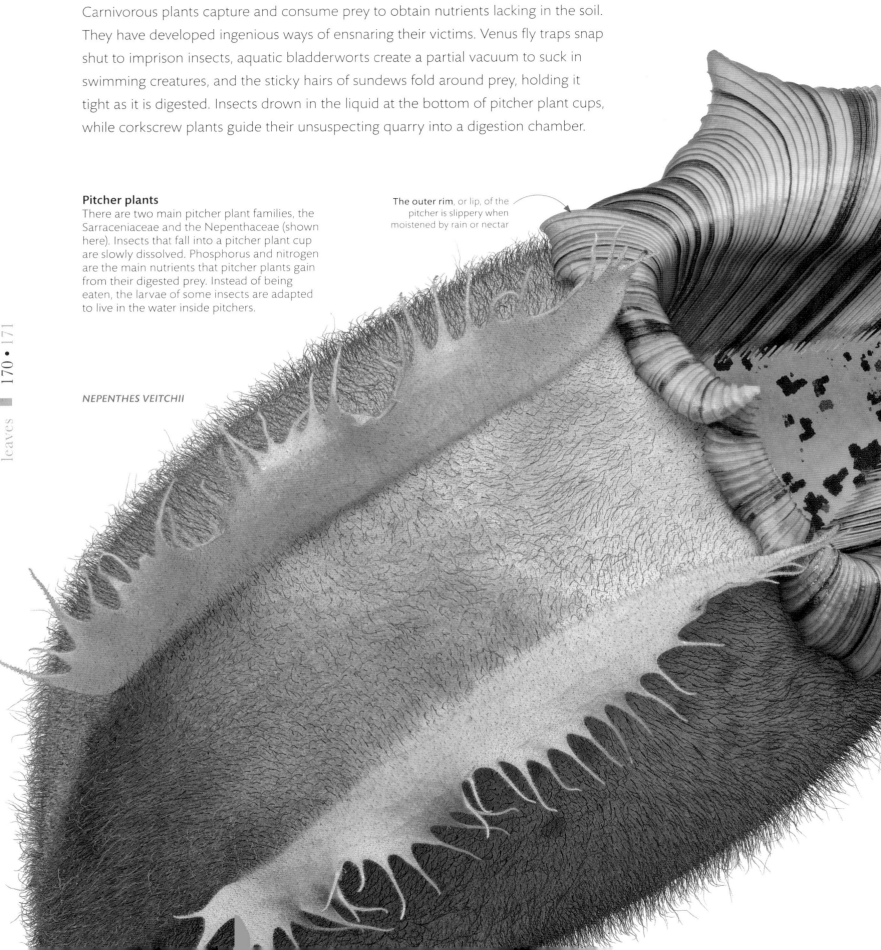

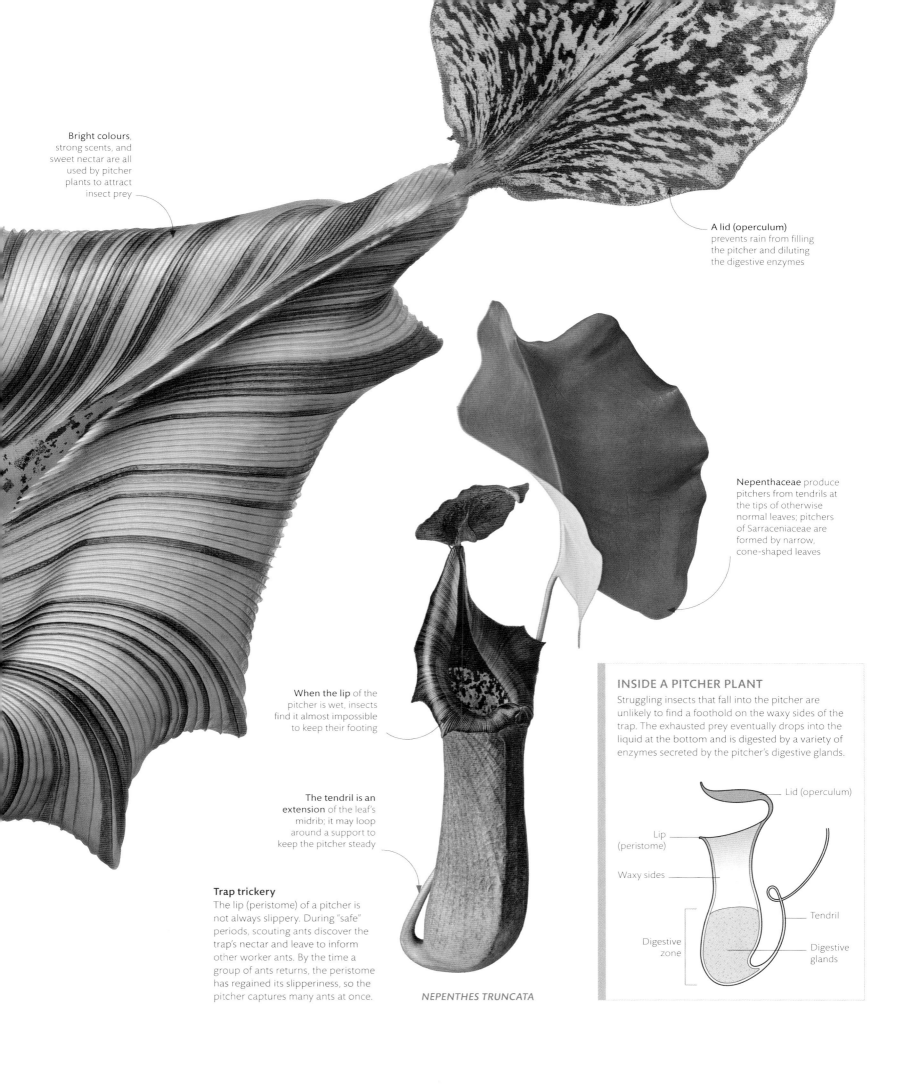

Bright colours, strong scents, and sweet nectar are all used by pitcher plants to attract insect prey

A lid (operculum) prevents rain from filling the pitcher and diluting the digestive enzymes

Nepenthaceae produce pitchers from tendrils at the tips of otherwise normal leaves; pitchers of Sarraceniaceae are formed by narrow, cone-shaped leaves

When the lip of the pitcher is wet, insects find it almost impossible to keep their footing

The tendril is an extension of the leaf's midrib; it may loop around a support to keep the pitcher steady

Trap trickery
The lip (peristome) of a pitcher is not always slippery. During "safe" periods, scouting ants discover the trap's nectar and leave to inform other worker ants. By the time a group of ants returns, the peristome has regained its slipperiness, so the pitcher captures many ants at once.

NEPENTHES TRUNCATA

INSIDE A PITCHER PLANT
Struggling insects that fall into the pitcher are unlikely to find a foothold on the waxy sides of the trap. The exhausted prey eventually drops into the liquid at the bottom and is digested by a variety of enzymes secreted by the pitcher's digestive glands.

Lid (operculum)

Lip (peristome)

Waxy sides

Digestive zone

Tendril

Digestive glands

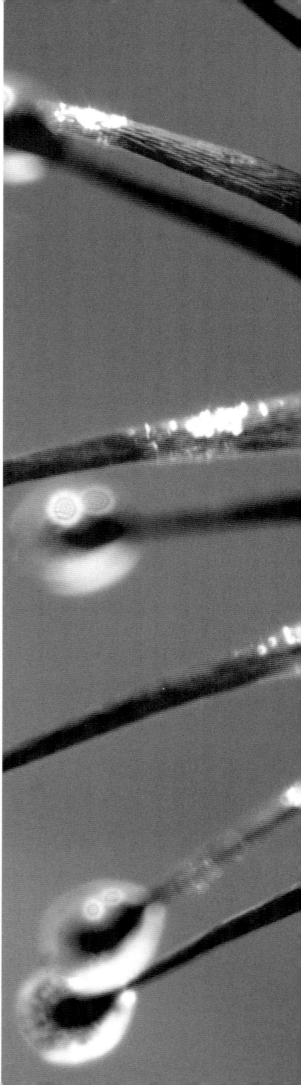

Drosera sp.

sundews

Armed with hundreds of sticky glands on each leaf, sundews are an insect's worst nightmare. Lured by the promise of sweet nectar or refreshing dew, visiting insects rapidly become ensnared by gluey drops of mucilage as the leaves curl around their flailing bodies and proceed to digest them.

The idea that there were plants with a carnivorous diet shocked many early naturalists. To Carl Linnaeus, it was an affront to the natural order of the divine plan. Not all views were so negative, though. Charles Darwin took great joy in studying sundews. Writing to a colleague, he went as far as to say: "at this present moment I care more about *Drosera* than the origin of all the species in the world."

The carnivorous appetite of sundews evolved in response to the nutrient-poor conditions of their typical habitats, which include bogs, swamps, marshes, and fens. With nitrogen in short supply, sundews have foregone symbiotic relationships with soil fungi and instead trap their own nitrogen-rich insect prey. Sensing the movement of insects arriving in search of nectar, their sticky glands swiftly close in on their prey. The more the victim struggles, the more securely it is held.

In most sundews, the process is completed by the leaf wrapping itself around the insect. The leaf then releases digestive fluids that break down the prey's body; other glands absorb the resulting juices.

All sundews produce attractive flowers far above their leaves, so as not to devour their pollinators. While sundews may seem exotic, they can be found growing on every continent except for Antarctica. Australia is the centre of diversity for this genus, boasting 50 per cent of all known species. Many species have become beloved, albeit curious, houseplants.

Deadly rosette
There are over 190 *Drosera* species, and they take many shapes and forms, from pygmy rosettes to tuberous climbing vines. The glistening, sticky droplets at the end of their "tentacles" give these plants the common name of sundews.

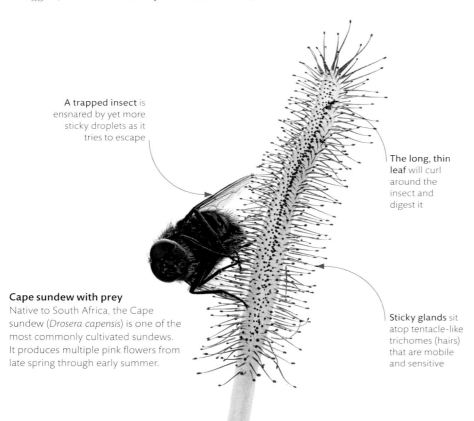

A trapped insect is ensnared by yet more sticky droplets as it tries to escape

The long, thin leaf will curl around the insect and digest it

Cape sundew with prey
Native to South Africa, the Cape sundew (*Drosera capensis*) is one of the most commonly cultivated sundews. It produces multiple pink flowers from late spring through early summer.

Sticky glands sit atop tentacle-like trichomes (hairs) that are mobile and sensitive

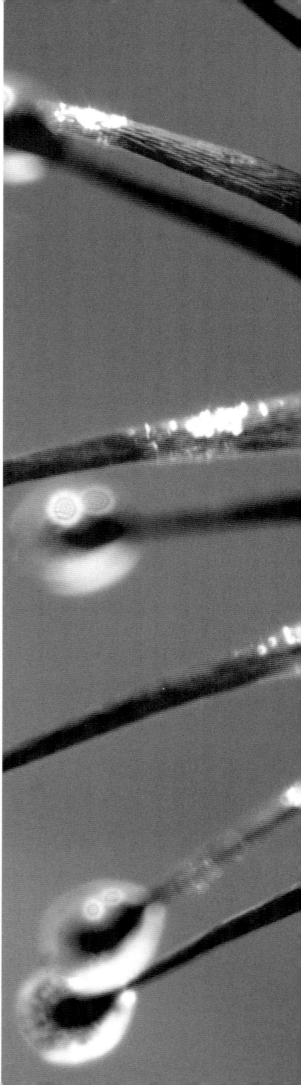

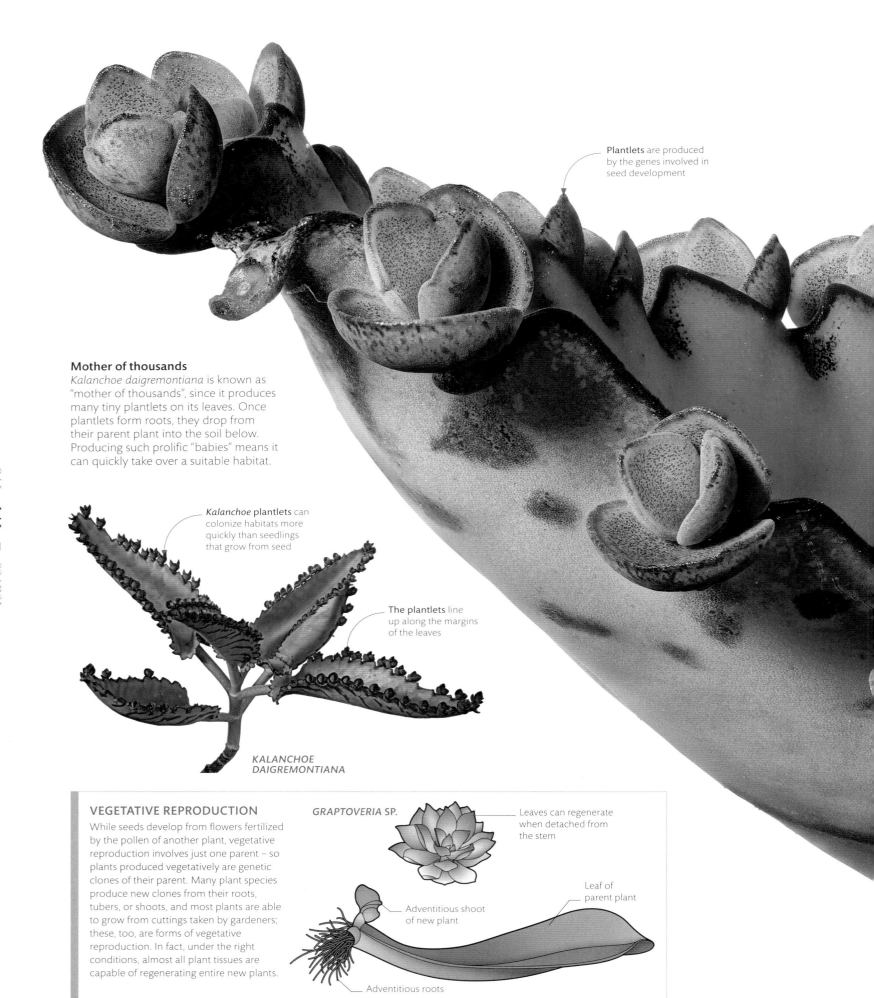

Plantlets are produced by the genes involved in seed development

Mother of thousands

Kalanchoe daigremontiana is known as "mother of thousands", since it produces many tiny plantlets on its leaves. Once plantlets form roots, they drop from their parent plant into the soil below. Producing such prolific "babies" means it can quickly take over a suitable habitat.

Kalanchoe **plantlets** can colonize habitats more quickly than seedlings that grow from seed

The plantlets line up along the margins of the leaves

KALANCHOE
DAIGREMONTIANA

VEGETATIVE REPRODUCTION

While seeds develop from flowers fertilized by the pollen of another plant, vegetative reproduction involves just one parent – so plants produced vegetatively are genetic clones of their parent. Many plant species produce new clones from their roots, tubers, or shoots, and most plants are able to grow from cuttings taken by gardeners; these, too, are forms of vegetative reproduction. In fact, under the right conditions, almost all plant tissues are capable of regenerating entire new plants.

GRAPTOVERIA SP.

Leaves can regenerate when detached from the stem

Adventitious shoot of new plant

Leaf of parent plant

Adventitious roots

leaves that reproduce

For plants, finding the perfect home is a matter of pure chance. Their seeds are blown, washed, or carried away from the parent plant, and could end up anywhere. When they find a suitable environment, some plants make copies of themselves (clones) to spread across their new home. Some of the most fascinating examples of this are plants that create tiny plantlets on their leaves. The plantlets are nurtured by the parent plant until they are big enough to survive on their own.

A plantlet receives nutrients from the parent plant until it detaches from the leaf margin

Roots and leaves form while the plantlet is still attached to its parent

NOURISHING BRACTS

The feathery bracts of the stinking passionflower (*Passiflora foetida*) exude a sticky substance that traps insects that might otherwise eat the flowers or fruit. The plant is protocarnivorous: it partially digests the trapped insects for nutrients.

PASSIFLORA FOETIDA

Sharp spines protect the developing flowers of the wild cardoon, but domesticated artichokes lack such protection

Bracts are modified leaves, but they do not look like the large, cleft foliage typical of cardoons

Spiky bracts

The cardoon (*Cynara cardunculus*) is a close relative of the globe artichoke. Its dramatic inflorescences are protected by an arrangement of thick, spiky bracts known as the involucre, which defends the soft developing floral tissues against both insect and mammalian herbivores. Each flower produces a single seed, or achene, topped by the hairy pappus (modified calyx) that aids dispersal by the wind.

protective bracts

Beneath many flowers and inflorescences are modified leaves called bracts. The function of bracts is twofold: some showy bracts mimic colourful petals to attract pollinators, while others provide a protective barrier around the developing flowers or fruit, shielding them from attack by herbivores or the elements. Hairy bracts can deflect wind and heat, while spines deter animals from taking a bite.

Hundreds of these **individual** purple flowers make up each inflorescence

The bracts contain antimicrobial compounds to defend the plant from fungi and bacteria

Silvery bracts, stalks, and leaves help cardoon plants to reflect heat and excess light

The fleshy base of each bract is edible, as is the artichoke heart hidden inside the shielding bracts

Leafy bracts resemble true leaves

LEAFY
Eucomis pole-evansii

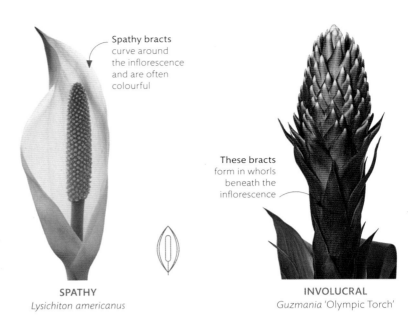

Spathy bracts curve around the inflorescence and are often colourful

SPATHY
Lysichiton americanus

These bracts form in whorls beneath the inflorescence

INVOLUCRAL
Guzmania 'Olympic Torch'

The lemma and palea bracts surround each grass flower inside the glumes

Each grass flower spikelet is protected by two scaly bracts called glumes

GLUME, LEMMA, AND PALEA
Melica nutans

Woody bracts fuse at base of inflorescence to protect the fruit

CUPULE
Quercus sp.

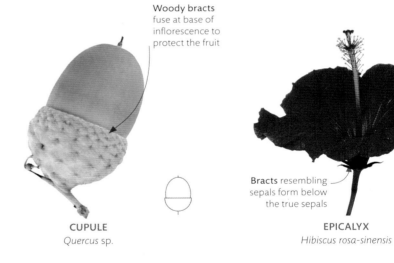

Bracts resembling sepals form below the true sepals

EPICALYX
Hibiscus rosa-sinensis

types of bract

Plants form many different types of bract, positioned beneath or around their flowers and inflorescences. Some bracts resemble leaves, while others look more like petals. Some bracts are deciduous and fall off before reproduction is complete, while others may be retained for the lifespan of the inflorescence to protect the developing fruit.

Brightly coloured bracts resemble petals to attract pollinators

PETALOID
Euphorbia pulcherrima

Each flower is protected by a tiny, stiff bract

SCALY
Humulus lupulus

Papery bracts
The petaloid bracts of *Bougainvillea* x *buttiana* are thin and papery. These showy structures surround the tiny flowers and attract pollinators while protecting the flowers. The bracts produce pigments called betalains, which make them bright pink.

New leaves
form at the top
of the plant

Showy white or
red petal-like
bracts surround
small yellow
flowers

Pairs of spines
are formed
from stipules
at the base
of each leaf

The leaf and stem sap
is toxic to many animals

Spines cover the
entire length of
the stem

leaves and spines

Many plants produce spines from leaves or parts of leaves, such as the
petiole (stalk) or the leafy outgrowths known as stipules. The main function
of spines is to defend the plant against herbivores. Some species, such as
cacti, have converted all their leaves into spines; with a reduced leaf surface
area, they lose less water through evaporation.

Crown of spines

Despite its common name, the crown of thorns plant (*Euphorbia milii*) actually produces spines rather than thorns (which grow from shoots). Spines, modified from the stipules at the base of the plant's leaves, help to defend its succulent stem against herbivores. The older leaves fall off as the plant grows, leaving a spine-covered stem with only a few new leaves at the top.

Spines help the crown of thorns to scramble over other plants

Leaf scars show the positions of the old leaves along the stem

The older part of the stem has no leaves

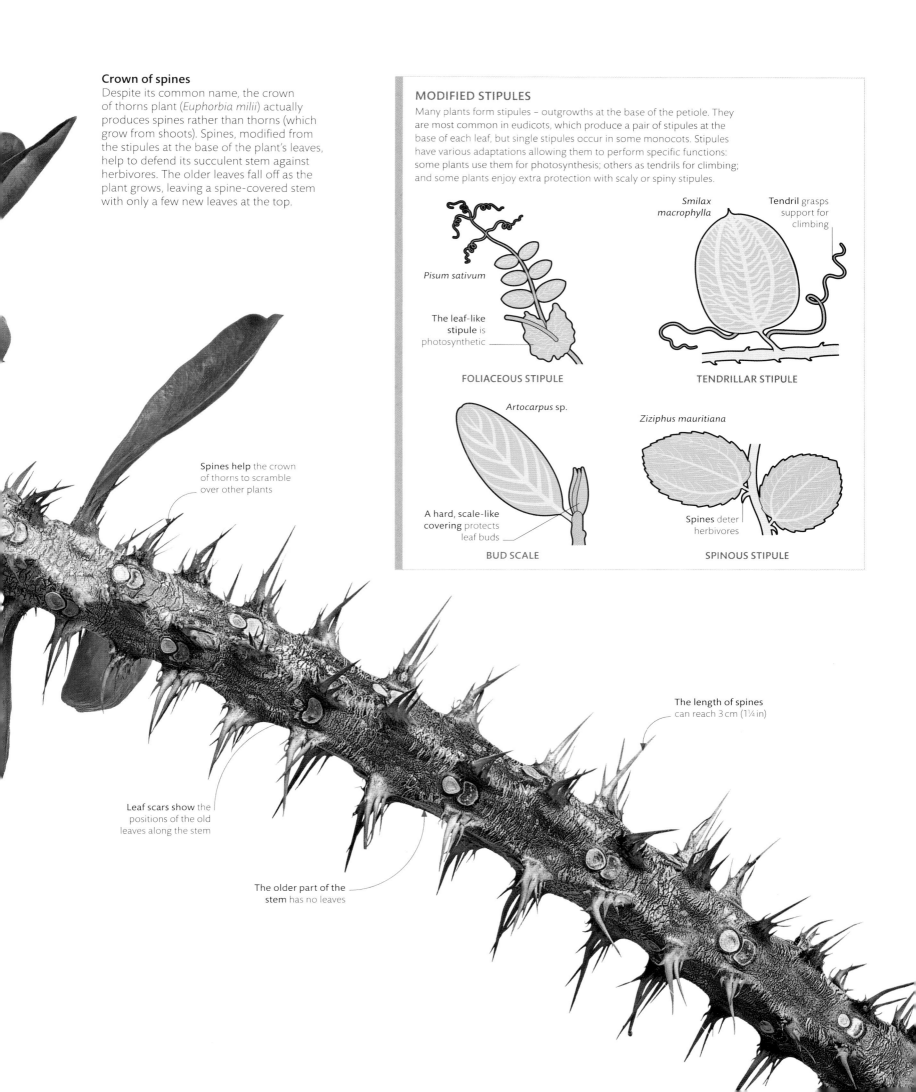

MODIFIED STIPULES

Many plants form stipules – outgrowths at the base of the petiole. They are most common in eudicots, which produce a pair of stipules at the base of each leaf, but single stipules occur in some monocots. Stipules have various adaptations allowing them to perform specific functions: some plants use them for photosynthesis; others as tendrils for climbing; and some plants enjoy extra protection with scaly or spiny stipules.

Pisum sativum

The leaf-like stipule is photosynthetic

FOLIACEOUS STIPULE

Smilax macrophylla

Tendril grasps support for climbing

TENDRILLAR STIPULE

Artocarpus sp.

A hard, scale-like covering protects leaf buds

BUD SCALE

Ziziphus mauritiana

Spines deter herbivores

SPINOUS STIPULE

The length of spines can reach 3 cm (1¼ in)

flowers

flower. the part of a plant from which its fruit or seed develops, consisting of reproductive organs (stamens and pistils) that are often surrounded by colourful petals and green sepals.

parts of a flower

Around 90 per cent of plants produce flowers, ranging from tiny, almost microscopic grass blooms to alien-looking giants that measure 1 m (3 ft) or more across. The most familiar are individual "complete" or "perfect" flowers – so-called because they contain both male and female reproductive organs within a single bloom.

COMPARING FLOWER STRUCTURES

Although they contain the same reproductive organs, monocots – such as lilies – differ from other flowering plants in the number and arrangement of these parts. Most monocots present petals, stamens, and ovaries in multiples of three; other flowering plants have petals and sepals in fours, fives, or indefinite numbers.

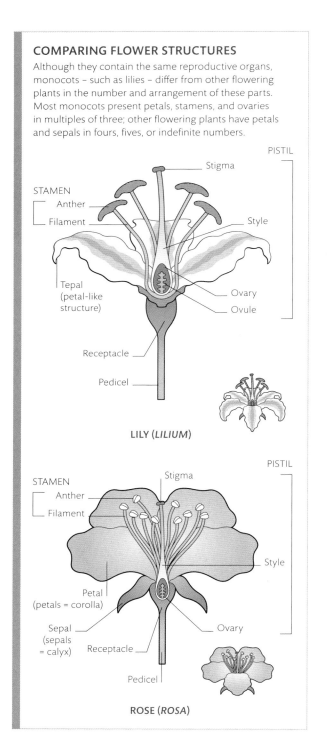

LILY (*LILIUM*)

ROSE (*ROSA*)

Simple flowers

A simple flower, such as this fuchsia, comprises male stamens and a female pistil surrounded by sepals and petals. Compare this with a compound ray and disk flower (see p.218).

A male anther, supported by the filament, produces pollen

Sepals enclose the flower head, peeling back when it blooms

FUCHSIA SP.

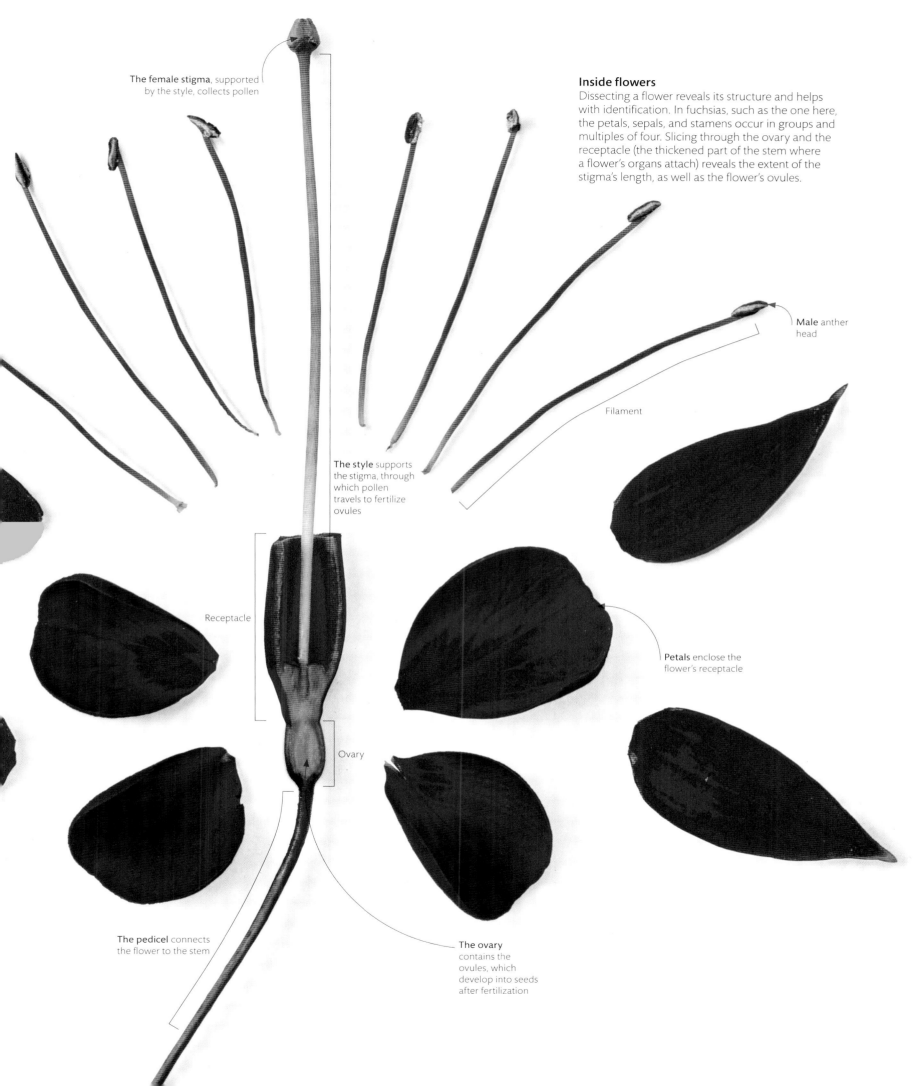

The female stigma, supported by the style, collects pollen

Inside flowers
Dissecting a flower reveals its structure and helps with identification. In fuchsias, such as the one here, the petals, sepals, and stamens occur in groups and multiples of four. Slicing through the ovary and the receptacle (the thickened part of the stem where a flower's organs attach) reveals the extent of the stigma's length, as well as the flower's ovules.

Male anther head

Filament

The style supports the stigma, through which pollen travels to fertilize ovules

Receptacle

Petals enclose the flower's receptacle

Ovary

The pedicel connects the flower to the stem

The ovary contains the ovules, which develop into seeds after fertilization

ancient flowers

Most flowering plants are defined as either monocots or eudicots, but some do not fit into either group. These are the so-called "primitive" species (basal angiosperms), that account for less than five per cent of flowering plants. Magnolias, laurels, peppers, and their relatives are thought to be the closest living descendants of the first flowering plants.

Early flower buds
Unlike many species of flowering plants, magnolia flower buds emerge after being encased in bracts rather than protective sepals.

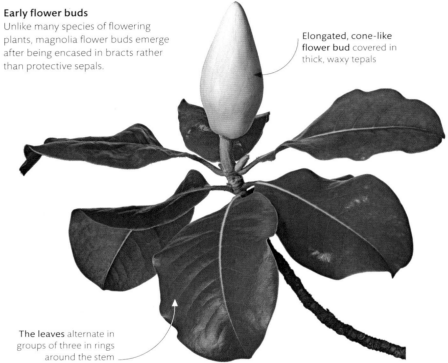

Elongated, cone-like flower bud covered in thick, waxy tepals

The leaves alternate in groups of three in rings around the stem

The outer flower consists of tepals – undifferentiated sepals and petals –arranged in whorls

ANCESTRAL BLOOM

Flowering plants first emerged around 247 million years ago, when water lily-like plants with simple flowers that produced both pollen and ovules appeared. Gradually, these diverged into trees and woody plants, herbaceous land plants, and aquatics. The ancient woody plants have survived more successfully than the others, and have since evolved into many different tree and shrub families. Japanese star anise (*Illicium anisatum*) is one example. With its reduced number of ovules merged into a star-like fruit, it is now thought to belong to one of the first groups of woody plants to evolve.

ILLICIUM

Family resemblance
Botanists believe that the earliest flower looked much like a southern magnolia (*Magnolia grandiflora*). Its flowers contain numerous, spirally arranged male and female organs, and produce cone-like seed receptacles when fertilized.

flower shapes

Botanists categorize flowers in various ways: noting how their reproductive organs are arranged, for example, or whether certain structures are present. Shape, however, is one of the most useful classifications. Noting whether or not a flower head is symmetrical, then examining the arrangement of petals (the corolla), provides the best starting point.

shape of corolla

Flat, wheel-shaped corolla at right angles to short tube

ROTATE
Lycianthes rantonnetii

Crown-shaped outgrowth of corolla

CORONATE
Narcissus 'Jetfire'

symmetry

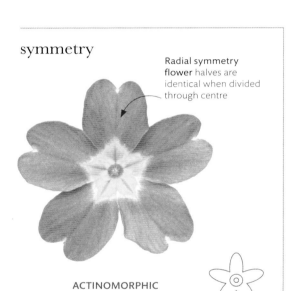

Radial symmetry flower halves are identical when divided through centre

ACTINOMORPHIC
Primula vulgaris

Corolla comprised of five petals, often overlapping

ROSATE
Rosa rubiginosa

Four petals at right angles to each other

CRUCIFORM
Cardamine pratensis

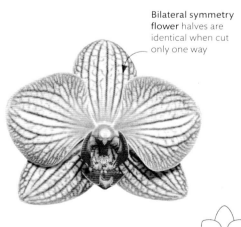

Bilateral symmetry flower halves are identical when cut only one way

ZYGOMORPHIC
Phalaenopsis sp.

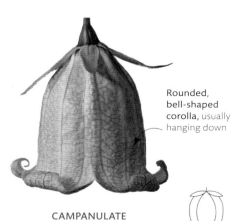

Rounded, bell-shaped corolla, usually hanging down

CAMPANULATE
Nesocodon mauritianus

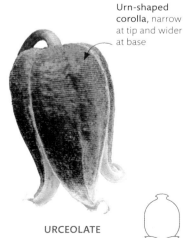

Urn-shaped corolla, narrow at tip and wider at base

URCEOLATE
Clematis viorna

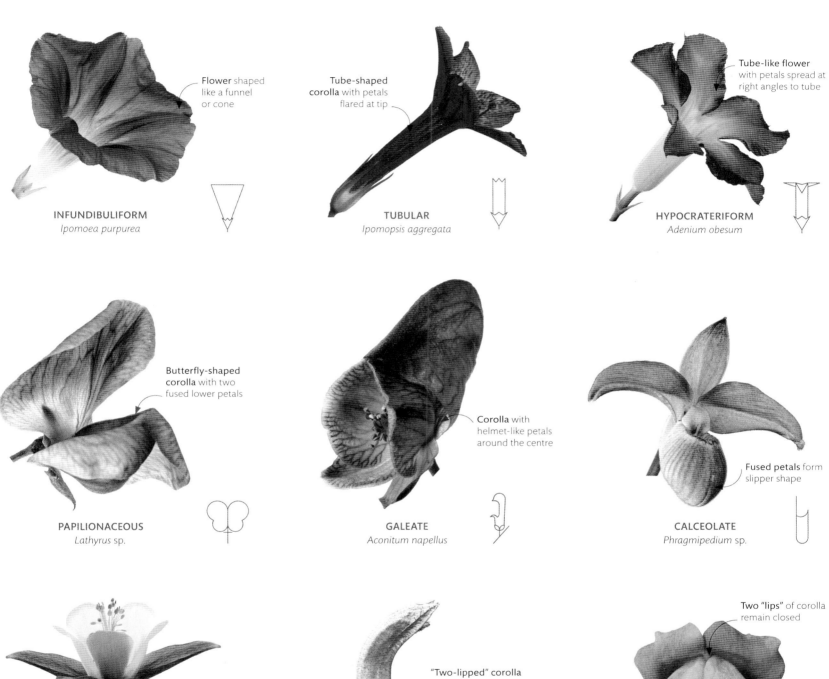

INFUNDIBULIFORM
Ipomoea purpurea

Flower shaped like a funnel or cone

TUBULAR
Ipomopsis aggregata

Tube-shaped corolla with petals flared at tip

HYPOCRATERIFORM
Adenium obesum

Tube-like flower with petals spread at right angles to tube

PAPILIONACEOUS
Lathyrus sp.

Butterfly-shaped corolla with two fused lower petals

GALEATE
Aconitum napellus

Corolla with helmet-like petals around the centre

CALCEOLATE
Phragmipedium sp.

Fused petals form slipper shape

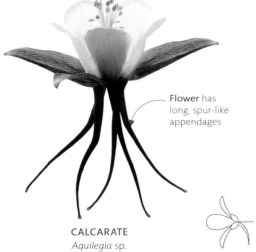

CALCARATE
Aquilegia sp.

Flower has long, spur-like appendages

BILABIATE
Brillantaisia lamium

"Two-lipped" corolla remains open

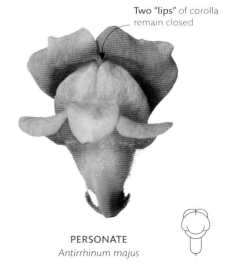

PERSONATE
Antirrhinum majus

Two "lips" of corolla remain closed

HERBACEOUS PERENNIALS

Botanically, any plant that flowers and sets seed year after year is called a perennial, from mighty oaks to humble geraniums. In horticulture, however, the term refers mainly to non-woody plants. The phrase "herbaceous perennial" denotes any soft-stemmed plant, such as a peony, whose stems die back completely in autumn or winter, in contrast to evergreen perennials, such as hellebores.

PAEONIA PEREGRINA

Purple stigmas protrude from the centre of the fully open bloom

The sepals unfurl, revealing green filaments topped by purple anthers

Nigella flowers bloom 10–12 weeks after seeds are sown

flower development

Flowers develop whenever a flowering plant is ready to reproduce. Life cycles range from weeks to years, depending on the species. Annuals germinate, flower, reproduce, and die within a year or less. Biennials spend a season developing from seed, rest over winter, and bloom the following spring, living for roughly two years. Perennials flower every year and live for three years or more. Some trees – woody perennials – can live for centuries.

The sepals begin to fold back, away from the maturing stamens and stigmas

The stigma withers and curls – it will form part of the multi-chambered seedhead

The anthers shrivel and die, eventually falling off

Life of a nigella flower
Annuals such as this *Nigella papillosa* 'African Bride', germinate, set seed, and die within a single growing season – no live root, stem, or leaf is left behind, and the plant only survives by its seed.

The longer tips of floret styles curl backwards as they mature

DAHLIA 'DAVID HOWARD'

GLOBE THISTLE (*ECHINOPS BANNATICUS* 'TAPLOW BLUE')

DAYLIGHT HOURS

Genetic changes triggered by exposure to sunlight make leaves produce a floral hormone called florigen, which "tells" a plant when it is time to flower. Plants make more or less florigen in response to daylight, and some species need more daylight than others.

8 DAYLIGHT HOURS

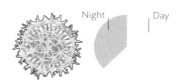

14 DAYLIGHT HOURS

ANY DAY LENGTH

Short-day plants need 8–10 hours of daylight, followed by 14–16 hours of darkness; if the darkness is interrupted at all, they do not bloom. Examples: poinsettias, dahlias, some soybeans.

Long-day plants need 14–16 hours of continuous daylight, followed by 8–10 hours of darkness; the daylight is more critical than the dark. Examples: globe thistle, lettuce, radishes.

Day-neutral plants are the most flexible flowering plants and bloom if exposed to anything from 5–24 hours of continuous daylight. Examples: sunflowers, tomatoes, and some peas.

flowers and seasons

Throughout the world, plants respond to the changing seasons. Regardless of which hemisphere they live in, most plants germinate, grow, and reproduce in spring and summer, when warm weather means that pollinators are plentiful. They release seed in late summer and autumn, when their growth slows down in preparation for winter. While these responses are partly a reaction to seasonal changes in temperature, the most crucial trigger is the changing amount of daylight, known as photoperiodism. Only plants at the equator experience days and nights of equal length all year round.

**EGLANTINE ROSE
(ROSA RUBIGINOSA)**

Blooming times
Plants need differing lengths of daylight to bloom, and their flowers open at different times of year. Globe thistle flowers open in mid-summer, whereas dahlias do not open until much later in the year. Some roses flower only in early summer, whereas others, such as most cultivated roses, may flower throughout the year.

人目之當呼紫玉昜生并題記

❝ Beautiful scenes throughout the year
are what you should remember. ❞

CHINESE PROVERB

plants in art

chinese painting

The strokes of ink and colour on silk and paper in Chinese flower painting have much in common with Chinese calligraphy. Scholars, trained from an early age in the art of handwriting, used calligraphic brushstrokes in their paintings. Flower painting was regarded as "silent poetry" and poetry as "painting with sound" and, over time, the two became combined in artworks that represent the song of nature.

Chinese flower painting originated from Buddhist banners decorated with flowers that were brought into China from India in the first century CE. The art form reached its peak in the Tang Dynasty (618–907) and has endured across the centuries.

The "four treasures" of traditional Chinese painting and calligraphy are ink stones, ink sticks, brushes, and paper. Artists use combinations of four basic techniques: outlines filled with colours, "boneless" washes, ink lines, and sketching ideas. The fine tip of a Chinese brush produces an infinite variety of strokes depending on which part of the brush is used and the pressure on paper or silk.

Plant subjects have a distinct character in the eye of the artist and a symbolism recognized throughout Chinese culture. In bird-and-flower painting, a distinct genre associated with the Taoist philosophy of harmony with nature, certain birds and flowers are paired for their symbolism: the crane and pine tree, for example, both represent longevity.

Flowers and poetry
A dainty sweep of magnolia blossoms is one of 12 ink and colour leaves inscribed with poems from an *Album of Fruit and Flowers*, the work of Qing Dynasty artist Chen Hongshou (1599–1652). Early flowering magnolias are celebrated in China as "the flower that welcomes spring". Legend has it that, at one time, only the Emperor was allowed to grow the tree.

Floral thoughts
Ming Dynasty artist Chen Chun (1483–1544) excelled in the study of ancient poetry, prose, and calligraphy, and brought this wealth of knowledge to his art. He regarded flower painting, such as this spontaneous collection of spring peach blossom and *Ziziphus jujuba*, as "idea writing".

Lateral buds mature after terminal buds

Pollen grains on the anthers are ready for transfer by insects, wind, or gravity

Open petals invite pollinators to land on the flowers, transferring pollen between blooms

Self-pollination

Plants such as St John's wort (*Hypericum pseudohenryi*) reproduce through the transfer of pollen from one flower to another on the same plant. This can also occur within the same flower. Both of these processes are described as self-pollination.

CROSS-POLLINATION

Cross-pollination occurs when pollen from an anther of one flower is transferred to the stigma of another flower on another plant of the same species. Once transferred, the pollen grains release a tube that pushes down the style into the ovary. This delivers male sperm cells to fertilize female egg cells within the ovary.

Pollen grains from anther of flower B

Pollen from flower A on stigma

Anther

Pollen tube

Style

Egg cell

Ovary

FLOWER A

FLOWER B

flower fertilization

The formation of flowers indicates that plants are ready to produce – or "set" – seed and pass on their genes. Fertilization occurs when sperm-bearing pollen is transferred to the flower's female reproductive organs, or pistil, where it fuses with the female reproductive cells (the ovules) to create seeds.

The anthers shrivel and the stamens droop after a flower is fertilized

A fertilized ovary changes colour, transforming into a red, ripening seedhead

The petals fold back and wither around a fertilized ovary, a sign the flower is starting to set seed

The stigma loses stickiness and turns brown on the ripening seedhead

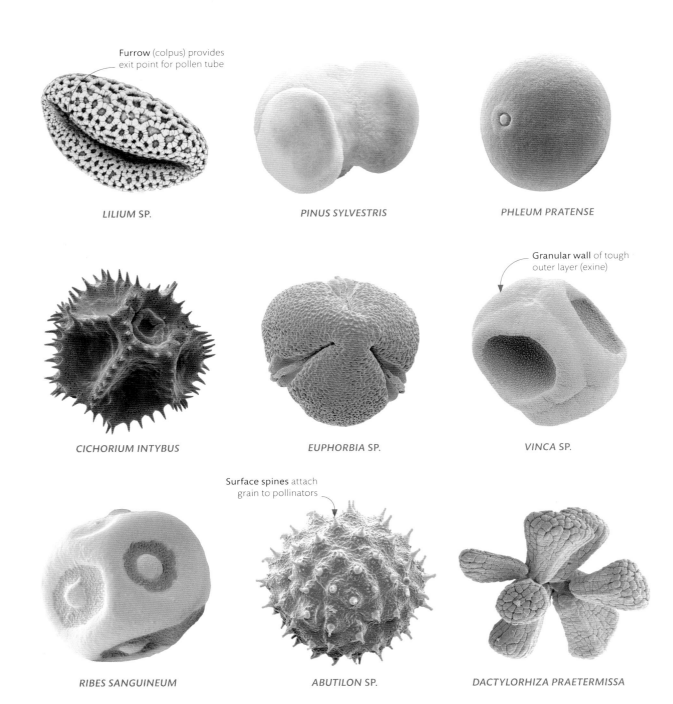

Furrow (colpus) provides exit point for pollen tube

LILIUM SP.

PINUS SYLVESTRIS

PHLEUM PRATENSE

Granular wall of tough outer layer (exine)

CICHORIUM INTYBUS

EUPHORBIA SP.

VINCA SP.

Surface spines attach grain to pollinators

RIBES SANGUINEUM

ABUTILON SP.

DACTYLORHIZA PRAETERMISSA

pollen grains

Although they look like dust to the human eye, pollen grains vary greatly in shape, size, and texture. Scanning electron microscopy reveals spheres, triangles, ovals, threads, and discs among other shapes. Grain surfaces can be smooth, sticky, spiny, striated, meshed, or grooved, and marked by pores or furrows.

Plentiful pollen
A honeybee can fill its leg baskets with around 15 mg (0.0005 oz) of pollen in a single foraging trip to flowers like this cactus.

Spatial considerations

The blooms of the Japanese lantern (*Hibiscus schizopetalus*) are located far apart. Hanging pendant-like from the ends of branches, they are easily seen by bird and insect pollinators.

A slender pedicel up to 15 cm (6 in) long supports each flower

Five long, lobed petals curve back, forming a globe-like shape

PIN AND THRUM

Some plants, such as the primrose, feature two incompatible flower types to reduce the possibility of self-pollination. The stigma in a "pin" flower is at the mouth of the floral tube; in a "thrum" flower it is located lower down the tube. Pollination between the two types is more likely to result in fertilization than pin-to-pin or thrum-to-thrum.

Petals

Stigma
Anthers
Filaments
Style
Ovary

PIN

Petals

Anthers
Filaments
Stigma
Style
Ovary

THRUM

The underside of the petals shows veining and blotching typical of many *Hibiscus* species

Green flower buds dangle from pedicels sprouting from leaf axils on the upper stem

Joint, where the pedicel connects to the peduncle

Self-avoidance
Most *Hibiscus* shed their pollen before the styles are receptive. They coat the legs and undersides of pollinators with pollen, which is carried to another bloom. When the styles become receptive, they curve up to receive pollen from another flower.

The staminal column extends from the centre of the flower and is twice as long as its petals

Pollen-loaded anther

Curved style

encouraging diversity

Many plants have evolved to favour pollination from other plants of the same species rather than from their own flowers, as cross-pollination usually results in stronger seeds and healthier plants that are more resistant to disease. One way in which plants can reduce the chance of self-pollination and increase the possibility of cross-pollination is by shedding their pollen before the styles are receptive, as is the case with the Japanese lantern.

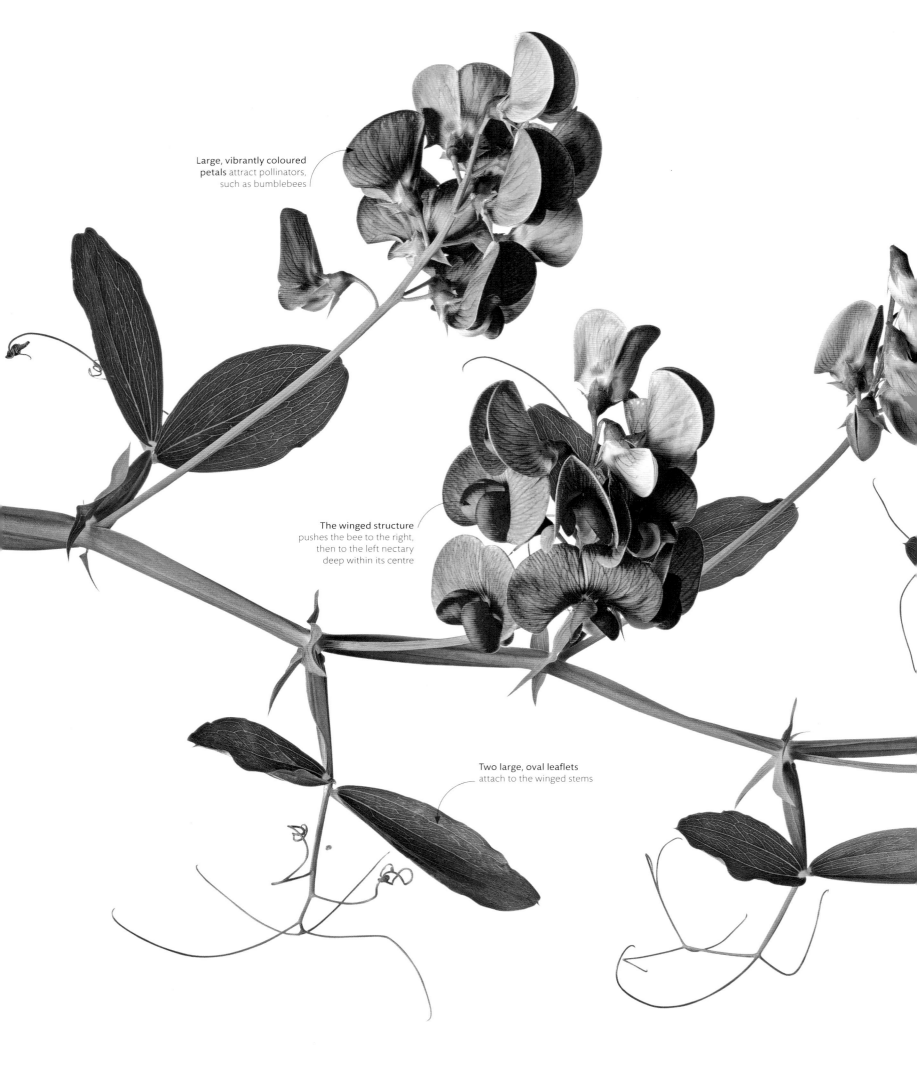

Large, vibrantly coloured
petals attract pollinators,
such as bumblebees

The winged structure
pushes the bee to the right,
then to the left nectary
deep within its centre

Two large, oval leaflets
attach to the winged stems

managing pollen

Flower shape can influence how plants release and receive pollen, and some plants have evolved blooms that make it difficult – or impossible – to pollinate themselves. Asymmetrical flowers such as pea blossom allow only the strongest insects to gain access; once inside, the internal structure ensures that pollen is received and dispersed in two separate stages – avoiding self-pollination.

Tendrils allow the plant to scramble up and over varied terrain

Two-stage pollination

Blooms of the everlasting pea (*Lathyrus latifolius*) have two nectaries. When a bee forces its way inside, the petals direct it first right, and then to the left, releasing the style, which has a hairy, brush-like area on which the pollen settles, below the stigma. The stigma touches the bee, collecting pollen from other flowers. When the bee moves to the other side, the brush pats it again, this time to transfer its own pollen for transport to other plants.

一種

一種
草斈鋸歯
あて紫色
淡紫辺の
物

千葉鋸歯あり
山朱色淡紫辺の物

Mount Fuji with Cherry Trees in Bloom
This polychrome woodcut by Hokusai, made around 1805, celebrates spring with a view of the snowy peak of Mount Fuji seen through cherry blossom and mist. The luxurious print, known as a *surimono*, is on thick paper and used metal pigments such as copper and silver powder.

plants in art

japanese woodblock prints

Chrysanthemum
At the height of his career as a landscape artist, Hokusai turned his attention to flowers, and produced a set of ten stylized woodcut prints known as the "Large Flowers". This detail from a larger work captures the detail of a chrysanthemum's many petals.

Honzo Zufu
Botanist Iwasaki Tsunemasa collected plants and seeds from the countryside and grew them in his garden so that he could record their fine detail in his artworks. This vibrant woodblock print of an opium poppy is taken from his *Honzo Zufu* (Botanical Atlas). The first four volumes were printed in 1828, and the entire work of 92 volumes was finally printed in 1921.

A mainstay of Japanese art for centuries, woodblock printing was at a peak of popularity in the 19th century. The prints were made with water-based inks that enhanced colour, glaze, and transparency, and their bold, simplified forms and subtle colours proved the perfect medium for capturing the beauty of the Japanese landscape and its native flora.

As the power of the last Shogunate declined and travel restrictions were lifted, Japan's botanists were drawn to Western scientific methods. Iwasaki Tsunemasa (1786–1842), a young shogun with a passion for the natural world, spent time with German scientist Philipp Franz von Siebold, who was based in the Dutch East Indies (now Indonesia). Iwasaki, also sometimes known as Kan-en, roamed the countryside, bringing back specimens to draw, paint, and name for his epic work, *Honzo Zufu*, a botanical atlas featuring 2,000 plants.

Perhaps the best-known artist of the Edo period is Katsushika Hokusai (1760–1849). He learned the art of woodblock printing, known as ukiyo-e (the art of the floating world), as a young apprentice and went on to excel in paint and print in every genre. Late in life he wrote: "At seventy-three, I began to grasp the structures of birds and beasts, insects and fish, and of the way plants grow. If I go on trying, I will surely understand them still better by the time I am eighty-six, so that by ninety I will have penetrated to their essential nature."

> " Illustrations must be made with all possible skill and accuracy. If they are not, how then will it be possible to differentiate plants which closely resemble each other? "

IWASAKI TSUNEMASA, PREFACE TO *HONZO ZUFU*

Male and female plants

Tiny, unisexual flowers are often smaller than their bisexual counterparts, but tend to mature faster. By producing a whole host of anther-bearing flowers, a male holly (*Ilex aquifolium*, shown left) increases its chances of attracting insects to carry pollen to female trees – which are the only ones to produce berries.

The ovary is clearly visible on female holly flowers

Anthers are empty

**FEMALE FLOWERS OF HOLLY
(*ILEX AQUIFOLIUM*)**

HOLLY BERRIES ON FEMALE TREE

unisexual plants

In the animal kingdom, separate, individual males and females are the norm when it comes to reproduction, but in the botanical world, when one plant bears flowers of only one sex, it faces great challenges. While such dioecious plants, which include many trees, avoid self-fertilization, they rely completely on pollen being transported successfully from male to female – often over quite a distance.

IMPERFECT FLOWER STRUCTURES

Flowers that contain only male or only female reproductive structures are classed as unisexual or "imperfect" and are self-incompatible. This means that they are unable to pollinate themselves in order to reproduce. When imperfect flowers of both sexes appear on the same plant – as in squashes and cucumbers – the plant is said to be monoecious. Whether existing on one plant or found on two separate individuals, however, male imperfect flowers can have many separate (free) stamens, or one central staminate structure comprised of fused anthers, filaments, or both.

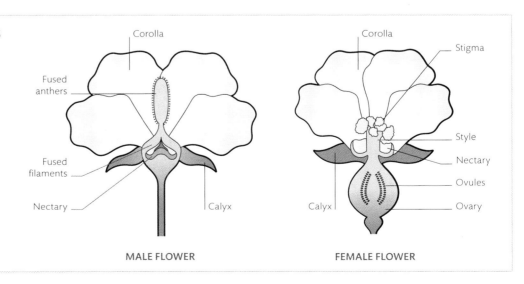

Corolla

Fused anthers

Fused filaments

Nectary

Calyx

Corolla

Stigma

Style

Nectary

Ovules

Ovary

Calyx

MALE FLOWER

FEMALE FLOWER

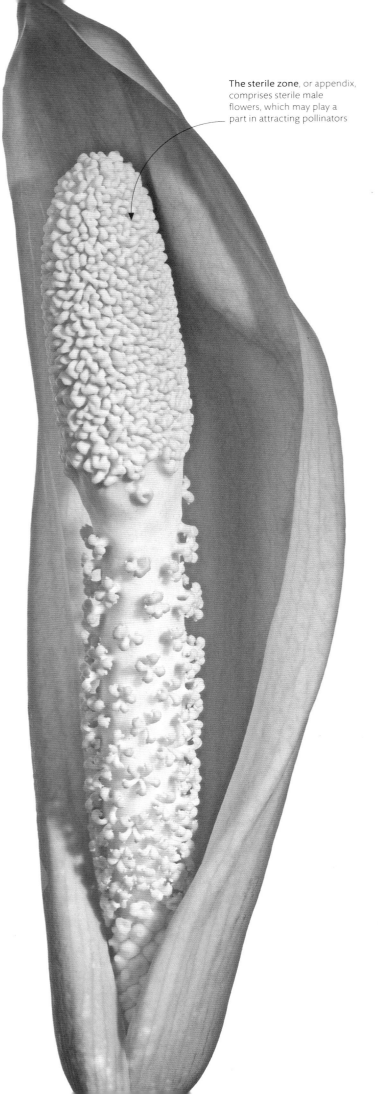

The sterile zone, or appendix, comprises sterile male flowers, which may play a part in attracting pollinators

incompatible
flowers

Many species avoid pollinating themselves by means of a careful arrangement of structures, in the case of bisexual flowers, or by having the unisexual male and female flowers on separate plants. Some inflorescences accomplish this even when they contain flowers of both sexes within one unit. Staggered maturity rates, leafy shields, and buffer zones ensure that these plants are largely cross-pollinated, creating a healthier genetic mix.

SEPARATION STRATEGY

Many arums, such as dumb canes (*Dieffenbachia* sp.), have sterile zones in the middle of the spadix, separating male and female flowers. These flowers never develop fully but perform the important role of helping to prevent pollen from fertile males reaching fertile females.

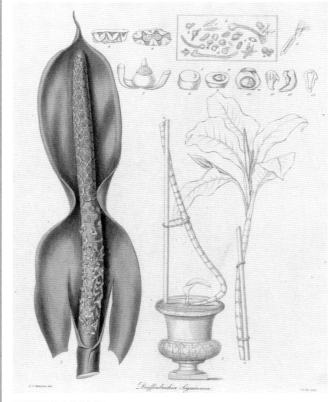

DIEFFENBACHIA SP.

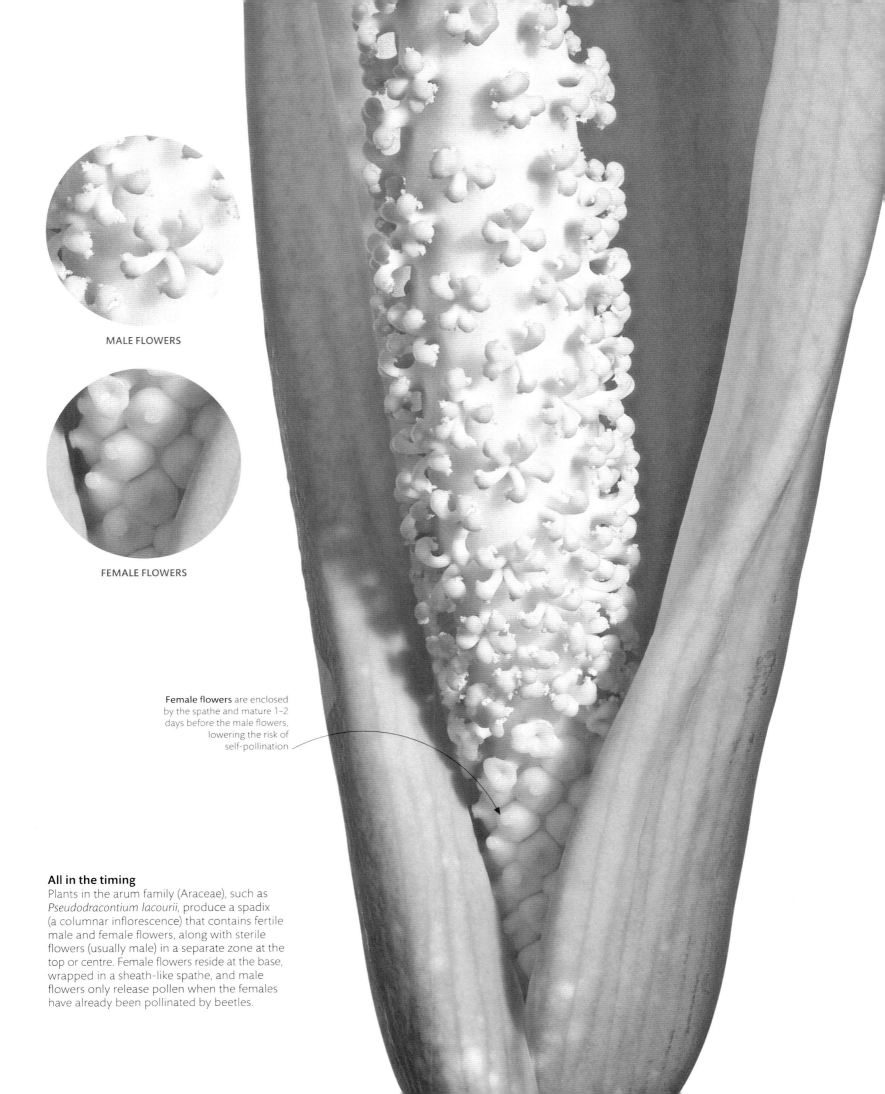

MALE FLOWERS

FEMALE FLOWERS

Female flowers are enclosed by the spathe and mature 1–2 days before the male flowers, lowering the risk of self-pollination

All in the timing
Plants in the arum family (Araceae), such as *Pseudodracontium lacourii*, produce a spadix (a columnar inflorescence) that contains fertile male and female flowers, along with sterile flowers (usually male) in a separate zone at the top or centre. Female flowers reside at the base, wrapped in a sheath-like spathe, and male flowers only release pollen when the females have already been pollinated by beetles.

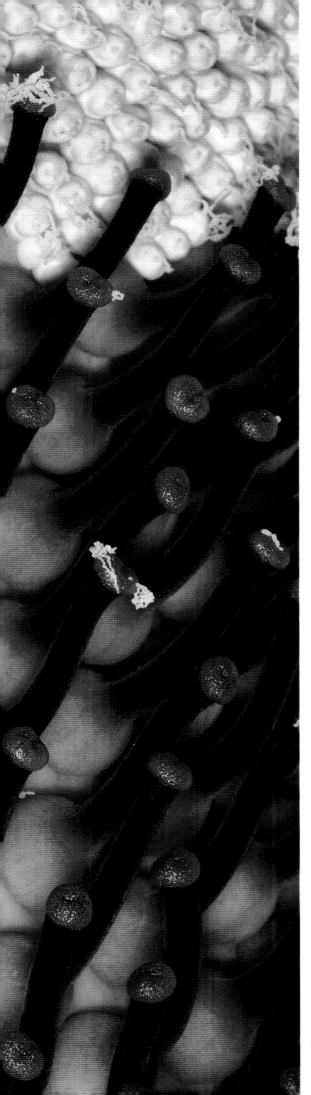

Amorphophallus titanum

the titan arum

This colossus of the Sumatran rainforest holds the record for the largest unbranched inflorescence in the world. The huge size of the inflorescence is matched by its powerful odour, which has been likened to that of rotting flesh – hence its nicknames of "corpse plant" and "carrion flower".

Appearances are deceptive with the titan arum: what looks like a 3 m (10 ft) tall flower is really an inflorescence known as a spadix, surrounded by a frilly tissue called a spathe. The flowers themselves are deep inside, at the base of the spadix.

This plant is remarkable not just for its inflorescence: it also generates its own heat. As the inflorescence matures, energy stored in a massive underground organ called a corm is used to warm the flowers to around 32°C (90°F). The corm, which weighs up to 50 kg (110 lb), is the largest known in the plant world. It is thought that producing all this heat helps the plant to disperse its foul scent throughout the dense rainforest to attract pollinators.

In essence, pollination for the titan arum relies on a ruse. The smell of rotting meat draws insects such as flesh flies and carrion beetles in search of a meal and a place to mate and lay their eggs. They find none of these rewards, and instead are doused with pollen. With luck, they fall for the trick again the next evening and pollinate another titan arum bloom.

The inflorescence itself only lasts 24 to 36 hours before it collapses. It takes so much effort to produce that individual plants can only bloom every three to ten years. When not in flower, the titan arum exists as a single, enormous leaf, roughly 4.5 m (15 ft) tall. As impressive as the titan arum is, this plant is at risk of becoming endangered if the deforestation of its habitat continues at the current rate.

Titanic bloom
The flowers of the titan arum are arranged in dense clusters, with male flowers on top (white) and female flowers at the bottom (red). The overall structure of the titan arum acts like a chimney, channelling the fetid smell up into the air column to spread it far and wide.

The fleshy, hollow spadix retains the heat produced underground

A frilly spathe surrounds and protects the flowers at the base of the spadix

Meaty allure
Hundreds of tiny flowers are hidden behind the titan arum's huge, petal-like spathe. The burgundy coloration of the spathe is thought to mimic the appearance of decomposing meat.

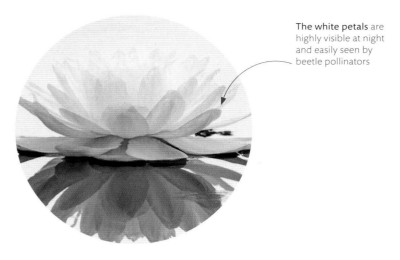

The white petals are highly visible at night and easily seen by beetle pollinators

Ladies first
Giant water lily flowers are relatively short-lived. They open at dusk, heating up and exuding a pineapple scent that is irresistible to scarab beetles of the *Cyclocephala* genus. The petals then close on the beetles, imprisoning them until the next evening.

flowers that change sex

Most flowering plants have hermaphroditic flowers containing both male and female parts. These separate sex organs usually mature alongside each other during the blooming cycle. In some hermaphrodites, however, the reproductive parts mature in such separate, distinct stages that the flowers effectively change sex – either from female to male (protogyny) or from male to female (protandry). The giant water lily (*Victoria amazonica*), which is pollinated by beetles trapped within its flowers, has flowers with female sex organs that mature before the male ones.

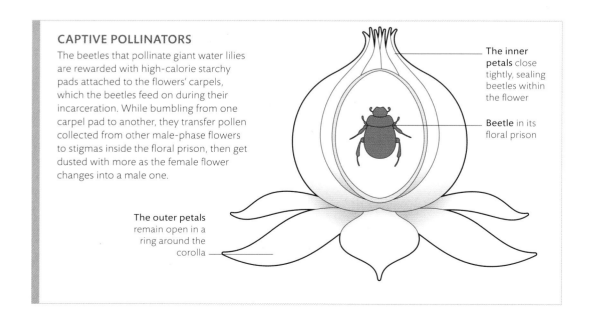

CAPTIVE POLLINATORS
The beetles that pollinate giant water lilies are rewarded with high-calorie starchy pads attached to the flowers' carpels, which the beetles feed on during their incarceration. While bumbling from one carpel pad to another, they transfer pollen collected from other male-phase flowers to stigmas inside the floral prison, then get dusted with more as the female flower changes into a male one.

The inner petals close tightly, sealing beetles within the flower

Beetle in its floral prison

The outer petals remain open in a ring around the corolla

Undergoing change
Changing from white to pinkish purple signals that a giant water lily is entering its male phase. All the beetles trapped inside the flower spend the day being dusted with pollen shed by the water lily's maturing anthers. When the flower opens on the second evening, the beetles fly away, seeking more female flowers. The flower then sinks below the water's surface, its task accomplished.

Bell-shaped
flowers blush
pink as they
mature

Drumstick allium
Allium sphaerocephalon features round
or egg-shaped umbels. In an umbel,
the inflorescence stalk, or peduncle,
has a wider, rounded end. This stalk
serves as a "platform" for numerous
flowers, whose stems, or pedicels,
are all the same length.

UMBEL

Anthers shed most of their pollen
within 1–2 days of opening, before
the stigma is fully developed. Once
mature, a stigma remains receptive
for several days, and may receive
pollen from lower, younger flowers

Pollen is transferred to
the topmost flowers as
bees move from younger
to older blooms

Petals change colour
to guide pollinators to
flowers that offer the
best rewards

Flowers at the base may
open up to two weeks later
than the topmost blooms

inflorescences

Many of the most striking floral displays occur in plants with an
inflorescence that produces numerous flowers on a single stem.
This is especially true of ornamental alliums. From a distance,
these members of the onion family give the impression of bearing
one large bloom, but a closer look reveals that each "bloom" consists
of numerous tiny flowers, or florets. If a single inflorescence's florets
open at different times over a period of days, or even weeks, and
each floret produces pollen capable of fertilizing its neighbours on
the same plant, the opportunity for self-pollination is immense.

Flourishing individuals
Many-flowered inflorescences cram multiple sources of
nectar and pollen into a small space, allowing pollinators
to feast while utilizing less energy in moving from flower to
flower. The blooms often mature at different rates, which
actively encourages pollinators to return; this strategy
promotes both self-pollination on the same inflorescence
as well as cross-pollination between different plants.

The topmost flowers often open first; as they age, the flowers may increase the amount of nectar they produce to encourage return visits from insects

types of inflorescence

Inflorescences are defined by how their flowers are arranged around main and lateral stalks, called peduncles and pedicels. Inflorescences can be either determinate or indeterminate. Determinate inflorescence stalks end in a single flower, while indeterminate ones terminate in a vegetative bud. Once a terminal flower bud forms, growth in that direction stops for determinate flowers. Indeterminate inflorescences, by contrast, continue to grow and can produce flowers at varying stages of maturity on the same inflorescence. Here are some examples of the many types.

determinate inflorescences

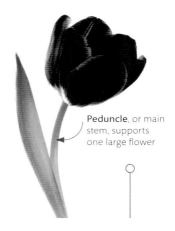

Peduncle, or main stem, supports one large flower

SOLITARY
Tulipa sp.

indeterminate inflorescences

Flowers grow around a central main peduncle

Flowers mature from the bottom up

RACEME
Delphinium sp.

Lateral branches support several flowers each

PANICLE
Cassia fistula

Peduncles resemble racemes and keep growing, but lateral branches end in flower buds, like cymes

THYRSE
Syringa sp.

Stalked flowers arranged alternately on peduncle

COMPOUND CORYMB
Hydrangea sp.

Stalked florets grow from a single point at peduncle's terminal end

UMBEL
Allium ursinum

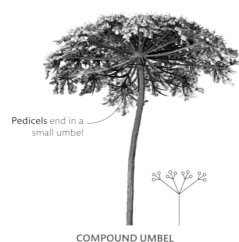

Pedicels end in a small umbel

COMPOUND UMBEL
Daucus carota

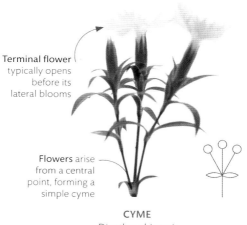

Terminal flower typically opens before its lateral blooms

Flowers arise from a central point, forming a simple cyme

CYME
Dianthus chinensis

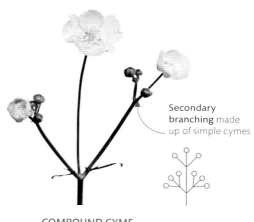

Secondary branching made up of simple cymes

COMPOUND CYME
Ranunculus acris

Alternate pedicels on each side, creating a zigzag flower pattern

SCORPIOID CYME
Iris sp.

Stalkless (sessile) flowers attached directly to the peduncle

SPIKE
Callistemon sp.

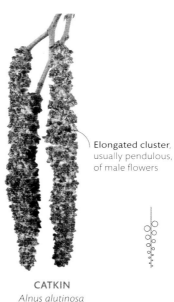

Elongated cluster, usually pendulous, of male flowers

CATKIN
Alnus glutinosa

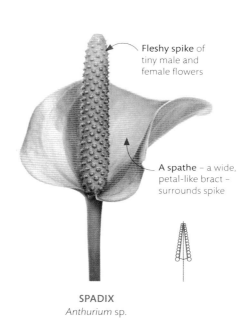

Fleshy spike of tiny male and female flowers

A spathe – a wide, petal-like bract – surrounds spike

SPADIX
Anthurium sp.

Densely packed florets attached directly to the end of the peduncle

CAPITULUM
Taraxacum sp.

Stalkless florets densely packed on disc-shaped head

CAPITULUM
Echinacea purpurea

Inflorescence arranged in a whorl

VERTICILLASTER
Stachys palustris

As ray flowers unfold, the strap-like shape of each individual petal is revealed

The whorl of bracts folds back, allowing the head to expand

Central disc flowers begin to colour and swell as the ray flowers elongate

ECHINACEA PURPUREA 'MAXIMA'

How composite flowers open
Composite blooms open from the outside in. Disc flowers expand and change colour as they mature, but only open once the surrounding rays have completely unfurled.

rays and discs

The daisy family (Asteraceae) is one of the largest groups of flowering plants. They have a distinct floral structure – each apparently "single" flower head is made up of tiny blooms, known as ray and disc florets. Some, like dandelions, are composed of petal-like rays; some, like thistles, contain only tube-like discs; while others, such as echinacea, have both ray and disc florets.

FLOWER STRUCTURE

While their petals (ligules) differ hugely, both ray and disc blooms have fused anthers that form a cylindrical shape, and a pappus – hair-like bristles that have replaced the sepals, or calyx, found in typical flowers – that aid seed dispersal.

Stigma

Anther cylinder

Corolla

Pappus

Ligule

Style

Ovary

DISC FLOWER

RAY FLOWER

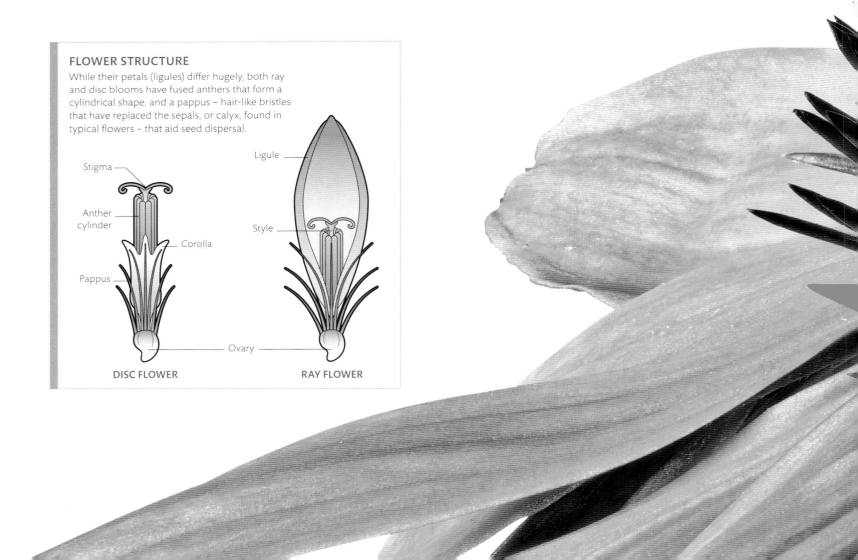

Many-flowered bloom
In echinacea heads, maturing disc flowers enlarge to form a rounded centre. Oval, pinkish ray flowers – which may be sterile, but help attract pollinators – flatten and flex backwards away from the discs. The tubular corollas blush orange-red, turning darker as they open to reveal forked, pollen-coated styles, each surrounded by five small, pointed edges.

Outer disc flowers turn from green at the base to red-orange towards the tip

Helianthus sp.

the sunflower

Sunflowers have been cultivated since at least 2600 BCE, not only for their bright yellow blooms that echo the sun, but also for their highly nutritious seeds. Originally native to the Americas, the sunflower has since spread across the globe.

Sunflowers are known for their habit of tracking the sun across the sky. Contrary to popular belief, this heliotropism only occurs as plants are developing. Both the leaves and the flower buds track the sun's path to maximize their exposure to its life-giving rays. Once the flowers have bloomed, this daily motion stops and the flowers generally orient themselves towards the east. In this way, they are able to take advantage of the sun's heat as soon as it rises over the horizon, which increases pollinator visits as well as the speed at which their seeds develop.

Big yellow inflorescence
Each flower in the inflorescence's centre, called a disc flower, produces a single seed. Most of the disc flowers shown here have yet to open. The more seeds a plant produces, the better the chances that it will have offspring the next year.

What looks like a single flower is really an inflorescence composed of many tiny flowers. The flowers of the inflorescence mature from the outside in, providing plenty of opportunity for pollinators throughout the blooming period.

The genus *Helianthus* has around 70 different sunflower species, mostly annuals or biennials. The most commonly encountered sunflower is *Helianthus annuus*, which has been selectively cultivated for centuries to grow a single outrageously large inflorescence atop a long, bristly stalk. Wild sunflowers look quite different, producing many branching stalks, each ending in a much smaller inflorescence.

Some sunflowers are allelopathic: they produce a chemical cocktail that inhibits other plants' growth. By poisoning plants around them, the sunflowers limit the competition that they face and increase the amount of seed they can produce.

Each "petal" is made up of the fused petals of an individual ray flower

False petals
The sunflower's bright yellow "petals" are actually sterile flowers, called ray flowers, that exist only to attract pollinators to the fertile disc flowers at the centre of the inflorescence.

Spring catkins

Among more noticeable wind-pollinated flowers are catkins, produced by trees such as hazel (*Corylus* spp.). Most catkins are formed of male flowers. The merest breeze releases clouds of pollen, which fertilizes the female flowers.

Two glumes, sheath-like lower or basal bracts, enclose the base of each spikelet

Each floret is held within two bracts: the lemma and palea

The palea is the short, inner bract

The longer bracts are called lemma

Spikelets have one or multiple florets

Shimmering seed heads

The northern sea oat (*Chasmanthium latifolium*) is a tall, clump-forming grass, native to the woodlands and inland waterways of central and eastern North America. It produces dangling inflorescences that are typical of many wind-pollinated plants. The plant is sometimes called "spangle grass" on account of its oat-like seedheads that shimmer in the sunlight.

The culm, or main stem, gives the large, drooping spikelets maximum exposure to breezes

Fertilized florets become more rigid before they release their seed

Tiny flower

wind-pollinated
flowers

Easily overlooked, most wind-pollinated flowers use breezes to carry their pollen, so they do not need to draw attention to themselves with showy petals. Many plants that use wind pollination, such as grasses and hazel catkins, hide their flowers in special bracts that protect the reproductive organs until they emerge for just long enough to be fertilized.

Seizing the day
From spring to autumn, male and female flowers emerge from green spikelets of beach-dwelling sea oats (*Uniola paniculata*). Florets open only once in the early morning, then quickly close.

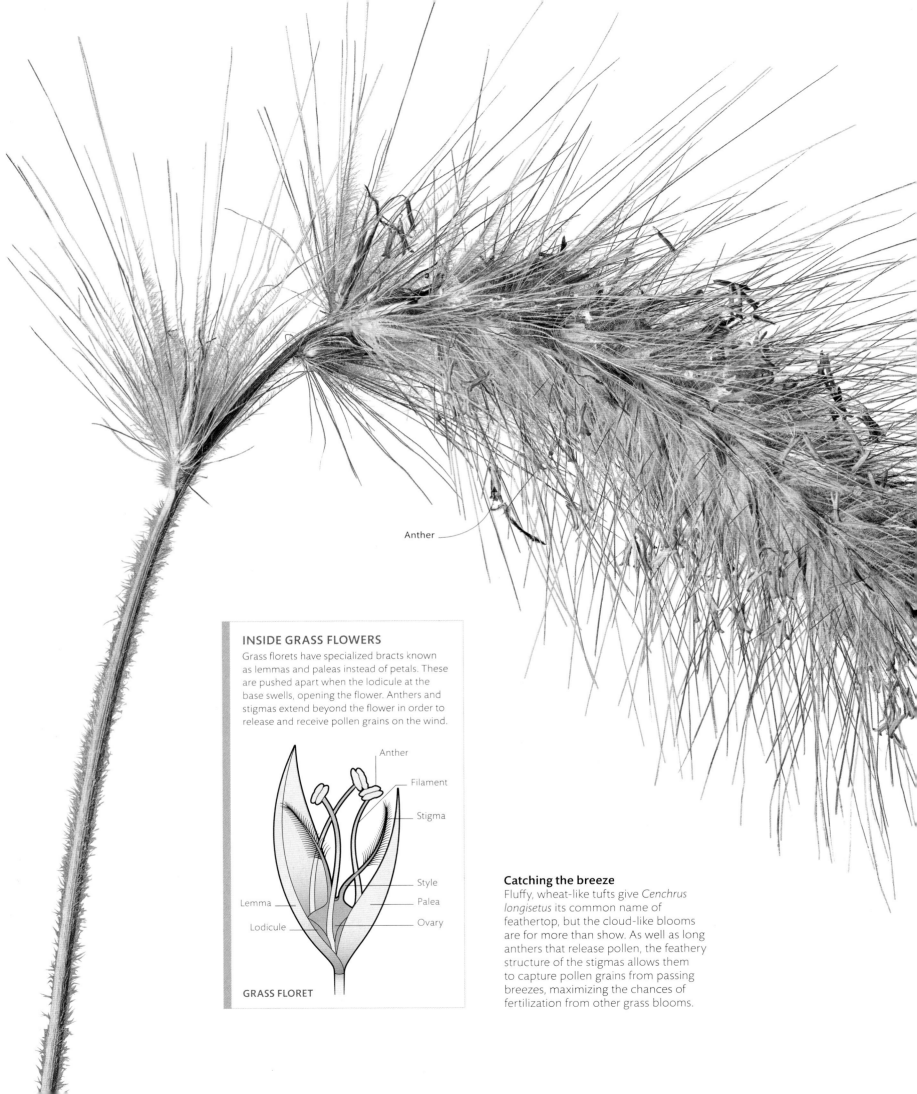

Anther

INSIDE GRASS FLOWERS

Grass florets have specialized bracts known as lemmas and paleas instead of petals. These are pushed apart when the lodicule at the base swells, opening the flower. Anthers and stigmas extend beyond the flower in order to release and receive pollen grains on the wind.

Anther

Filament

Stigma

Style

Lemma

Palea

Lodicule

Ovary

GRASS FLORET

Catching the breeze

Fluffy, wheat-like tufts give *Cenchrus longisetus* its common name of feathertop, but the cloud-like blooms are for more than show. As well as long anthers that release pollen, the feathery structure of the stigmas allows them to capture pollen grains from passing breezes, maximizing the chances of fertilization from other grass blooms.

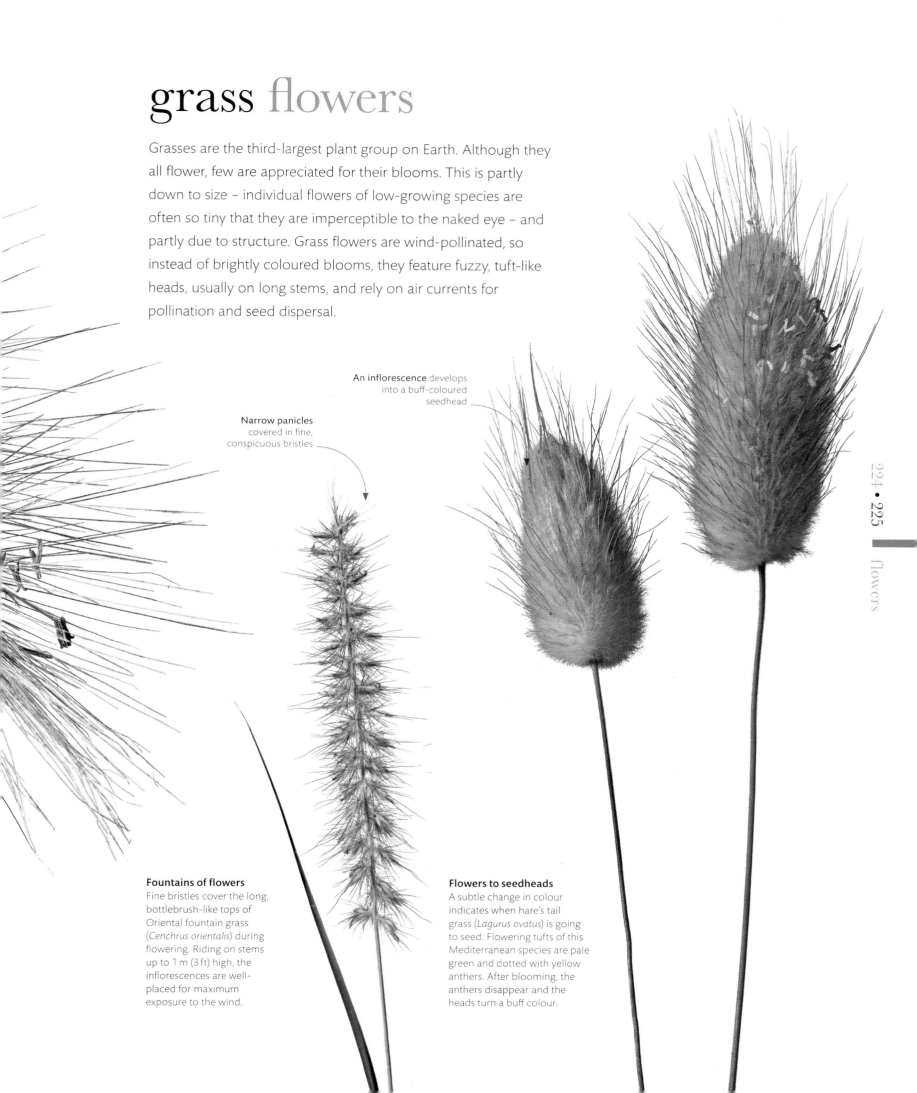

grass flowers

Grasses are the third-largest plant group on Earth. Although they all flower, few are appreciated for their blooms. This is partly down to size – individual flowers of low-growing species are often so tiny that they are imperceptible to the naked eye – and partly due to structure. Grass flowers are wind-pollinated, so instead of brightly coloured blooms, they feature fuzzy, tuft-like heads, usually on long stems, and rely on air currents for pollination and seed dispersal.

An inflorescence develops into a buff-coloured seedhead

Narrow panicles covered in fine, conspicuous bristles

Fountains of flowers
Fine bristles cover the long, bottlebrush-like tops of Oriental fountain grass (*Cenchrus orientalis*) during flowering. Riding on stems up to 1 m (3 ft) high, the inflorescences are well-placed for maximum exposure to the wind.

Flowers to seedheads
A subtle change in colour indicates when hare's tail grass (*Lagurus ovatus*) is going to seed. Flowering tufts of this Mediterranean species are pale green and dotted with yellow anthers. After blooming, the anthers disappear and the heads turn a buff colour.

Nectar-seeking geckos clamber over flowers, collecting sticky pollen on their bodies as they go

NESOCODON MAURITIANUS

Vital relationship
In certain habitats, reptiles serve as key pollinators. In Mauritius, the ornate day gecko laps nectar from the blossoms of the ox tree (*Polyscias maraisiana*), and pollinates this endangered species in the process.

The bell-shaped bloom hangs downwards, so only adept climbers can reach the sweet rewards

Pale blue petals ensure the blood-red nectaries stand out to attract the plant's gecko pollinators

flowers and nectar

Nectar is a plant's ultimate bribe. The sweet, sticky liquid, produced by glands called nectaries, attracts a wide range of pollinators. While nectaries also arise on stems, leaves, and buds, they are most closely associated with flowers, where nectar serves as a reward for pollination. Sugars such as sucrose, glucose, and fructose are the main constituents of nectar, alongside trace amounts of amino acids and other substances. The type, volume, and even colour of the nectar a flower makes differ from species to species – tailored to the tastes of the animals that pollinate them.

Unusual colour
Most nectar is colourless, relying on scent to attract pollinators, but the blue Mauritius bellflower (*Nesocodon mauritianus*) is an exception. To increase its chances of reproduction, it releases scarlet fluid from blood-red nectaries. The colour red is very attractive to the ornate day geckos that pollinate these rare bellflowers in their rocky habitat.

FLORAL NECTARIES
Within a flower, the three most common positions for nectaries are at the base of the ovary, the base of the stamens (specifically the filament), and at the petal bases. All these locations require pollinators to push past the reproductive parts of the flower to access the nectar – a design that fosters pollination. However, nectar-exuding glands may also occur on other parts of the ovary, the anthers, stamens, pistils, stigmas, and on petal tissue.

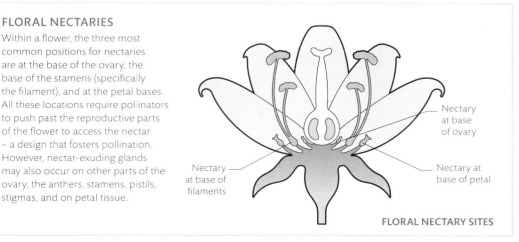

Nectary at base of ovary

Nectary at base of filaments

Nectary at base of petal

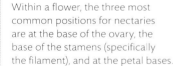

FLORAL NECTARY SITES

Petal nectaries
surround centre

HELLEBORE FLOWER

Specialized nectaries
The nectaries of the golden columbine
(*Aquilegia chrysantha*) are highly specialized.
Hidden in chambers at the tips of petal spurs,
they are only accessible to certain long-tongued
hawkmoth species. Hellebore nectaries are
located in cones that can only be reached by
bumblebees, which have long proboscises.

NECTARIES

Nectary positions vary according to which pollinators are available. The exposed nectaries of autumn-flowering ivy (*Hedera*) exude "pools" of nectar favoured by short-tongued wasps and flies. In early spring, nectar collects in the bases of hellebores' conical petal nectaries, while in summer-flowering monkshood (*Aconitum*), two large nectaries are hidden within the flower's hood. Both attract bumblebees.

Surface secretion

HEDERA

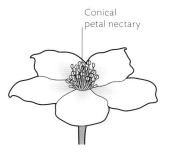

Conical petal nectary

HELLEBORUS

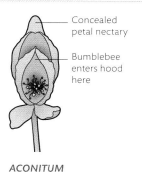

Concealed petal nectary

Bumblebee enters hood here

ACONITUM

storing nectar

As flowering plants evolved, nectaries (glands that secrete nectar) became concentrated within reproductive flower heads, and floral colours and scents developed to advertise the presence of nectar. Although many nectaries formed on the surface of flowers to attract a range of pollinators, others evolved to become accessible only to certain animals.

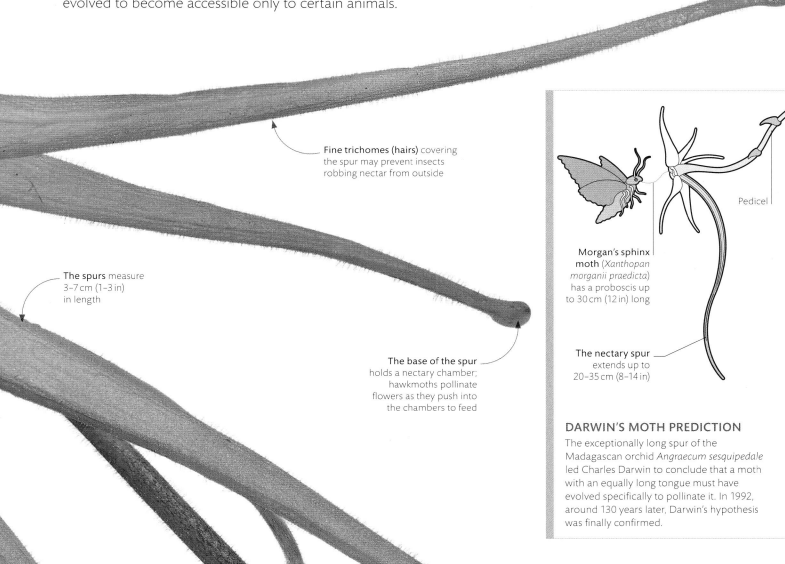

Fine trichomes (hairs) covering the spur may prevent insects robbing nectar from outside

The spurs measure 3–7 cm (1–3 in) in length

The base of the spur holds a nectary chamber; hawkmoths pollinate flowers as they push into the chambers to feed

Morgan's sphinx moth (*Xanthopan morganii praedicta*) has a proboscis up to 30 cm (12 in) long

Pedicel

The nectary spur extends up to 20–35 cm (8–14 in)

DARWIN'S MOTH PREDICTION

The exceptionally long spur of the Madagascan orchid *Angraecum sesquipedale* led Charles Darwin to conclude that a moth with an equally long tongue must have evolved specifically to pollinate it. In 1992, around 130 years later, Darwin's hypothesis was finally confirmed.

BOWL-SHAPED BLOOMS

Wide, open flowers such as poppies are ideal for flying insects, particularly bees, which can land easily on individual, bowl-shaped heads. The ability to alight easily on the exposed floral parts mean that pollinators expend less energy. This benefits the flowers because it means that a greater number of poppies are visited by pollinators – and are therefore fertilized – in a shorter amount of time.

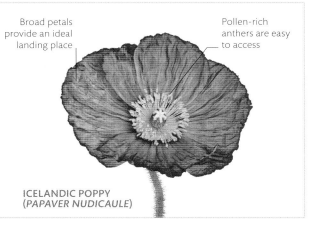

Broad petals provide an ideal landing place

Pollen-rich anthers are easy to access

ICELANDIC POPPY
(*PAPAVER NUDICAULE*)

designed for visitors

Floral colour and scent undoubtedly play crucial roles in attracting the attention of pollinators, but the shape of a flower can also determine just who those pollinators might be. Birds – except hummingbirds – need perches, while insects such as bees need landing platforms. Providing these features not only encourages the right visitors – it also enables some flowers to guide the right pollinators into their reproductive structures at the right moment.

Landing pads
Landing stages take many forms. The spongy domes of plume thistles (*Cirsium rivulare*) help butterflies and bees to grip onto them.

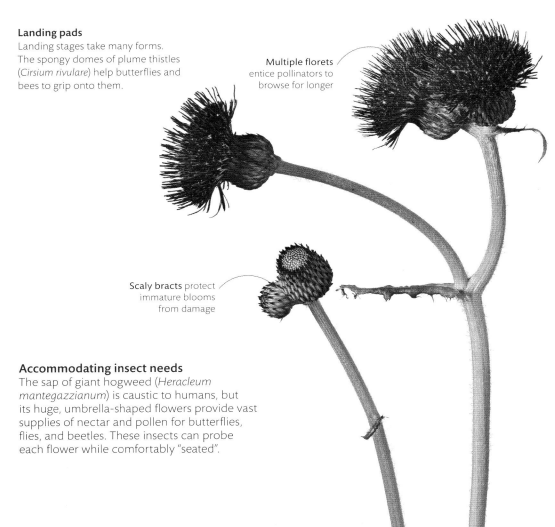

Multiple florets
entice pollinators to browse for longer

Scaly bracts protect immature blooms from damage

Accommodating insect needs
The sap of giant hogweed (*Heracleum mantegazzianum*) is caustic to humans, but its huge, umbrella-shaped flowers provide vast supplies of nectar and pollen for butterflies, flies, and beetles. These insects can probe each flower while comfortably "seated".

The bell shape forces bees to close their wings as they move along the tunnel towards the store of nectar

Anthers release pollen when bees brush along the ceiling as they move up the tunnel

Nectar is found inside the narrow end of the "bell"

Guard hairs deter small insects from entering the flower

Tailored for bumblebees
Foxgloves attract bumblebees with their combination of bee-specific colours, nectar guides, and "landing area". The flowers open sequentially from the bottom to the top, and the anthers mature before the stigmas.

DIGITALIS PURPUREA

precision pollination

Some plants have evolved to restrict access to certain pollinators. Everything about common foxglove (*Digitalis purpurea*) flowers seems designed to accommodate just one insect – the bumblebee. Elongated, enclosed blooms produce nectar that only long-tongued insects can reach, while guard hairs prevent smaller bees from entering. Bumblebees, however, are just the right size, shape, and weight to both plunder and pollinate these tubular flowers.

MUTUAL ADVANTAGE

The reproductive organs of a foxglove are positioned so that bumblebees are forced beneath them as they look for the nectary at the base of the ovary. The tight fit of the tubular flower ensures that bees are covered with pollen from the anthers and from grains dropped at the bottom of the flower. This maximizes the exchange of pollen as bees move between the flowers of different foxglove plants.

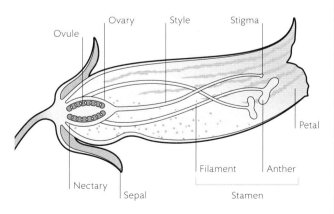

INSIDE A FOXGLOVE FLOWER

buzz pollination

Around 20,000 species attract pollinators using high-protein pollen as a lure. To increase the chance of passing on their genes, most of them, including potatoes and tomatoes, have distinctive structures that allow only specific insects access to their pollen. The insects, in turn, have evolved ingenious ways of collecting their reward. In buzz pollination, for example, certain bees transfer vibrations to flower heads to shake out pollen grains.

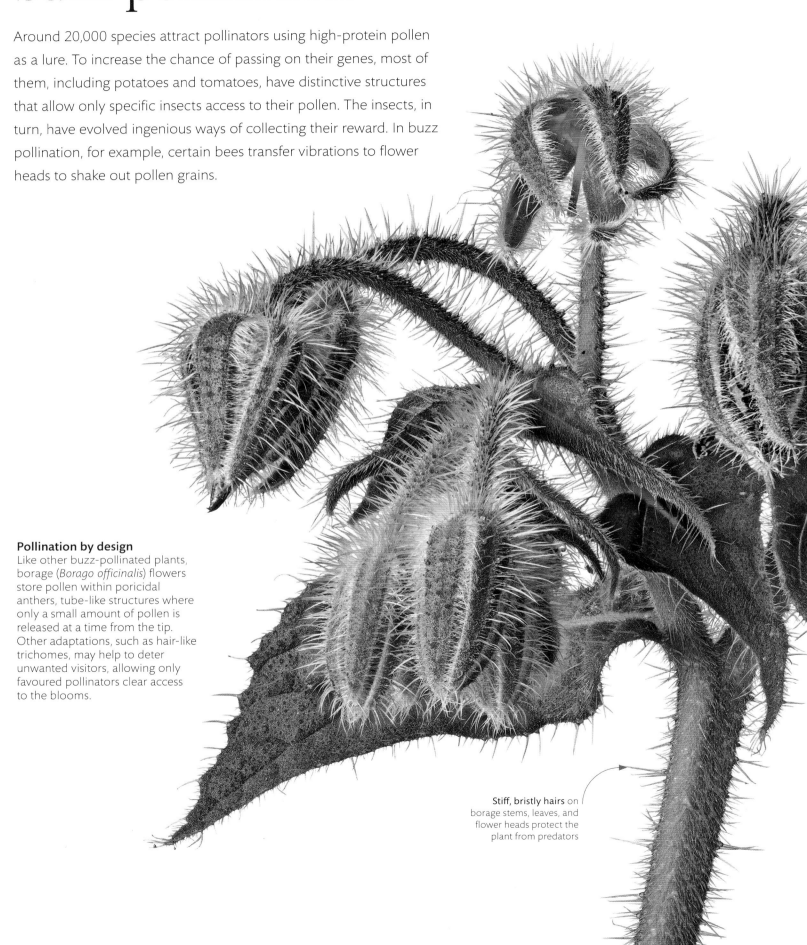

Pollination by design
Like other buzz-pollinated plants, borage (*Borago officinalis*) flowers store pollen within poricidal anthers, tube-like structures where only a small amount of pollen is released at a time from the tip. Other adaptations, such as hair-like trichomes, may help to deter unwanted visitors, allowing only favoured pollinators clear access to the blooms.

Stiff, bristly hairs on borage stems, leaves, and flower heads protect the plant from predators

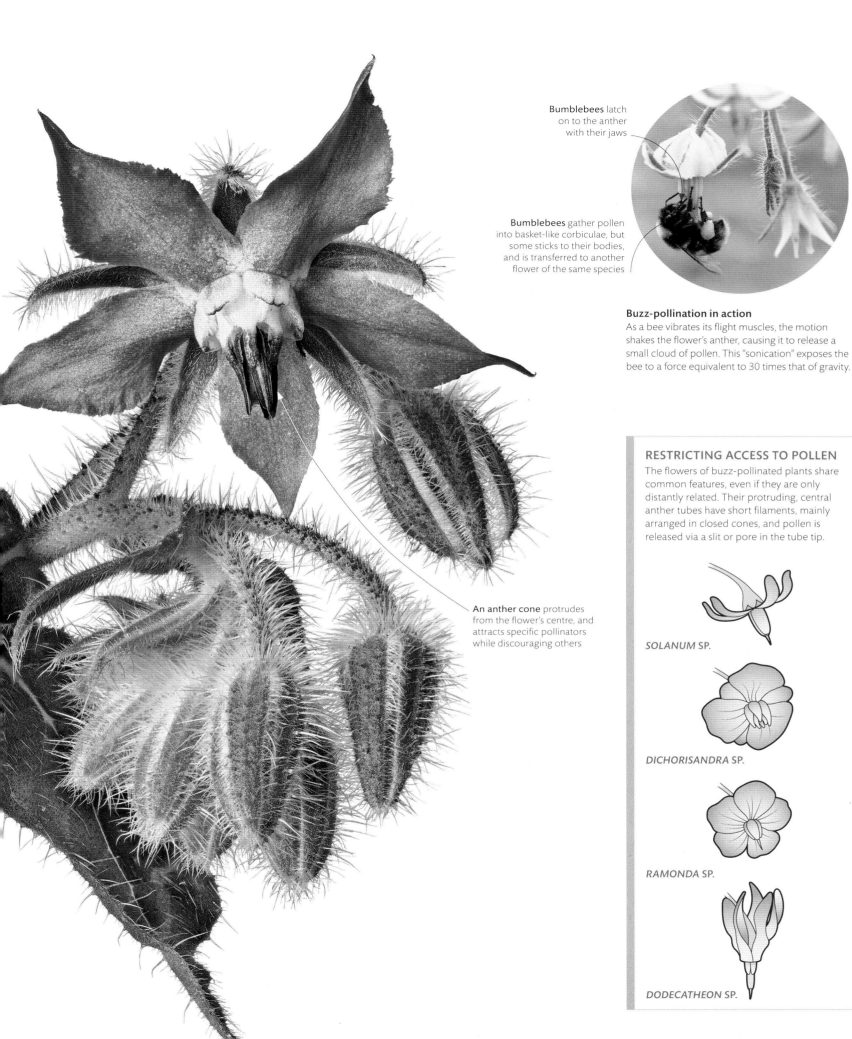

Bumblebees latch on to the anther with their jaws

Bumblebees gather pollen into basket-like corbiculae, but some sticks to their bodies, and is transferred to another flower of the same species

Buzz-pollination in action
As a bee vibrates its flight muscles, the motion shakes the flower's anther, causing it to release a small cloud of pollen. This "sonication" exposes the bee to a force equivalent to 30 times that of gravity.

An anther cone protrudes from the flower's centre, and attracts specific pollinators while discouraging others

RESTRICTING ACCESS TO POLLEN
The flowers of buzz-pollinated plants share common features, even if they are only distantly related. Their protruding, central anther tubes have short filaments, mainly arranged in closed cones, and pollen is released via a slit or pore in the tube tip.

SOLANUM SP.

DICHORISANDRA SP.

RAMONDA SP.

DODECATHEON SP.

Jimson Weed, 1936
The blossoms of the jimson weed (*Datura wrightii*),
a common desert plant that thrives on roadsides
and wasteland in the US, are magnified in Georgia
O'Keeffe's largest floral canvas. O'Keeffe was immensely
fond of the plant, in spite of the toxicity of its seeds, and
captured its pinwheel growth habit in this exuberant
composition, charged with energy and movement.

> " … they will be surprised into taking time to look at it – I will make even busy New Yorkers take time to see what I see of flowers. "

GEORGIA O'KEEFFE

plants in art

radical visions

As modernist artists sought new techniques to reflect the urban mechanized landscape that surrounded them in the early 20th century, the faithful and realistic representation of the natural world fell by the wayside. Their radical approach rejected centuries of representational art to focus on abstraction, introspection, and a return to primitivism. Over time, these new freedoms inspired artists to create modern works with an intense response to nature.

US artist Georgia O'Keeffe was already an early embracer of the modernist movement when she produced the floral paintings in the 1920s and '30s that were to become her most iconic works. These large-scale close-ups on giant canvases played with perspective, drawing viewers into the heart of a rose, or their gaze upwards to a towering canna lily.

Modernism was shot through with Freudian psychology and O'Keeffe's folds of petals and open blooms were given burdensome interpretations of female eroticism. It was not what she intended. Her rationale was to capture the fine detail of plants and write it large, so that people could witness the everyday miracle of plant beauty at a glance.

Flower power resurfaced in the 1960s as a hippie symbol of peace; blooms with simple shapes, bold patterns, and bright colours became fixtures of contemporary design. Although Andy Warhol's playful postmodern series of silkscreen prints entitled *Flowers* was a startling departure from his works based on commercial brands and popular culture, it was a perfect fit for the times.

Postmodernist art
In his pop art series *Flowers*, Andy Warhol experimented with swatches of colours for silkscreen prints based on a photograph of four hibiscus flowers. The colours of the flowers change from yellow, red, and blue, to pink and orange, or all white, against a backdrop of a bed of grass.

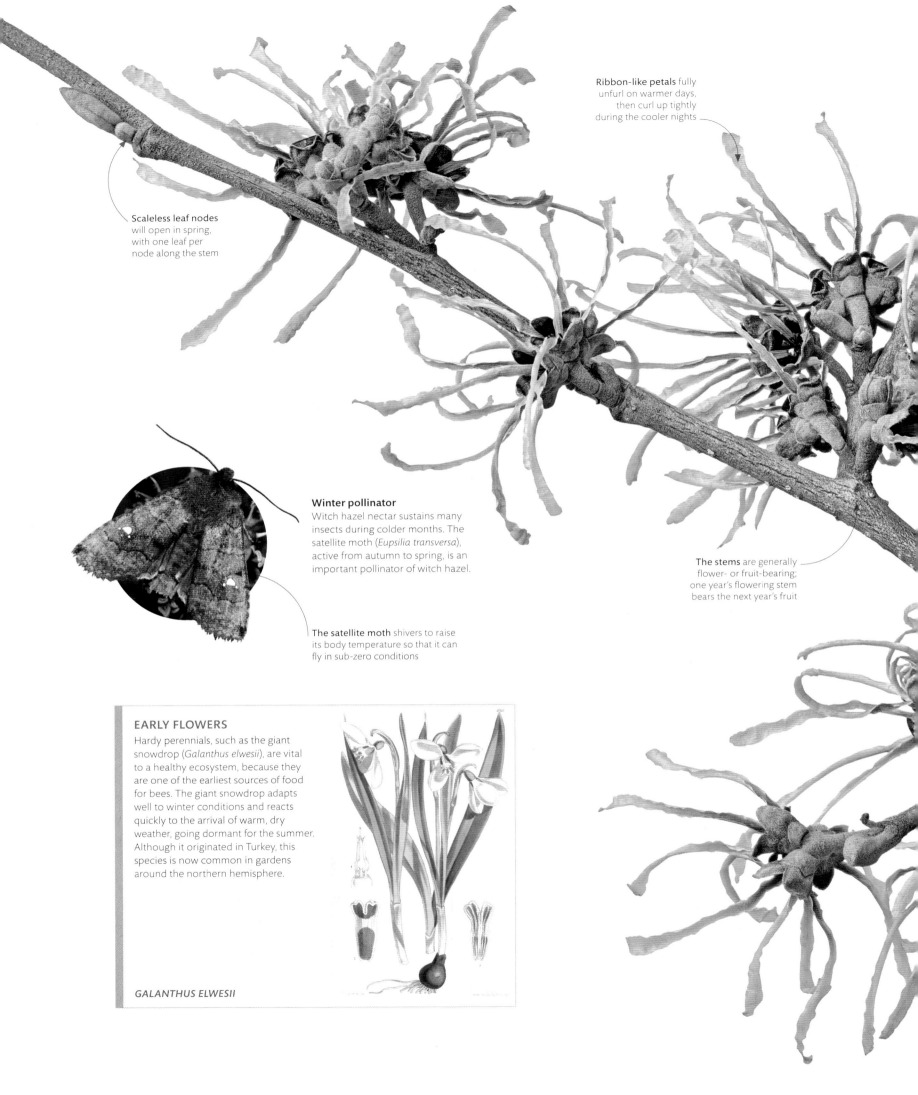

Ribbon-like petals fully unfurl on warmer days, then curl up tightly during the cooler nights

Scaleless leaf nodes will open in spring, with one leaf per node along the stem

The stems are generally flower- or fruit-bearing; one year's flowering stem bears the next year's fruit

Winter pollinator

Witch hazel nectar sustains many insects during colder months. The satellite moth (*Eupsilia transversa*), active from autumn to spring, is an important pollinator of witch hazel.

The satellite moth shivers to raise its body temperature so that it can fly in sub-zero conditions

EARLY FLOWERS

Hardy perennials, such as the giant snowdrop (*Galanthus elwesii*), are vital to a healthy ecosystem, because they are one of the earliest sources of food for bees. The giant snowdrop adapts well to winter conditions and reacts quickly to the arrival of warm, dry weather, going dormant for the summer. Although it originated in Turkey, this species is now common in gardens around the northern hemisphere.

GALANTHUS ELWESII

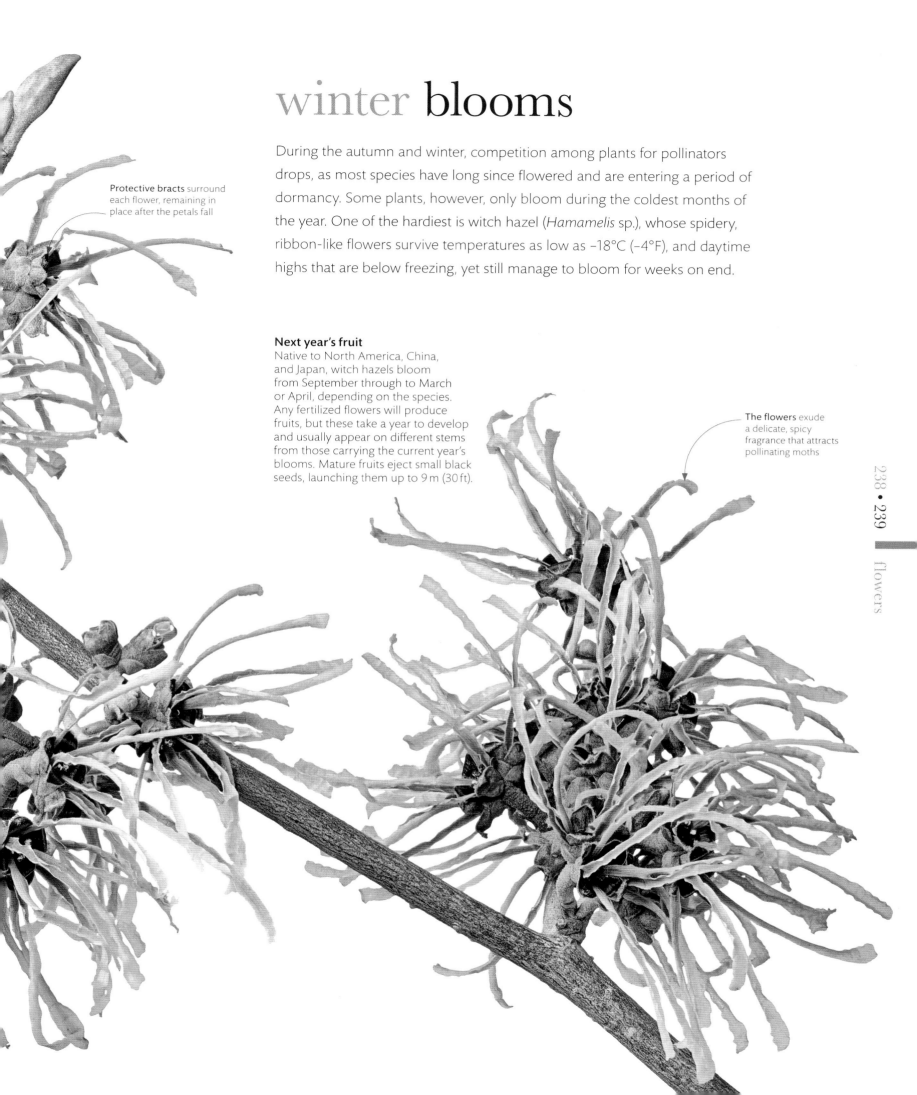

winter blooms

During the autumn and winter, competition among plants for pollinators drops, as most species have long since flowered and are entering a period of dormancy. Some plants, however, only bloom during the coldest months of the year. One of the hardiest is witch hazel (*Hamamelis* sp.), whose spidery, ribbon-like flowers survive temperatures as low as −18°C (−4°F), and daytime highs that are below freezing, yet still manage to bloom for weeks on end.

Protective bracts surround each flower, remaining in place after the petals fall

Next year's fruit
Native to North America, China, and Japan, witch hazels bloom from September through to March or April, depending on the species. Any fertilized flowers will produce fruits, but these take a year to develop and usually appear on different stems from those carrying the current year's blooms. Mature fruits eject small black seeds, launching them up to 9 m (30 ft).

The flowers exude a delicate, spicy fragrance that attracts pollinating moths

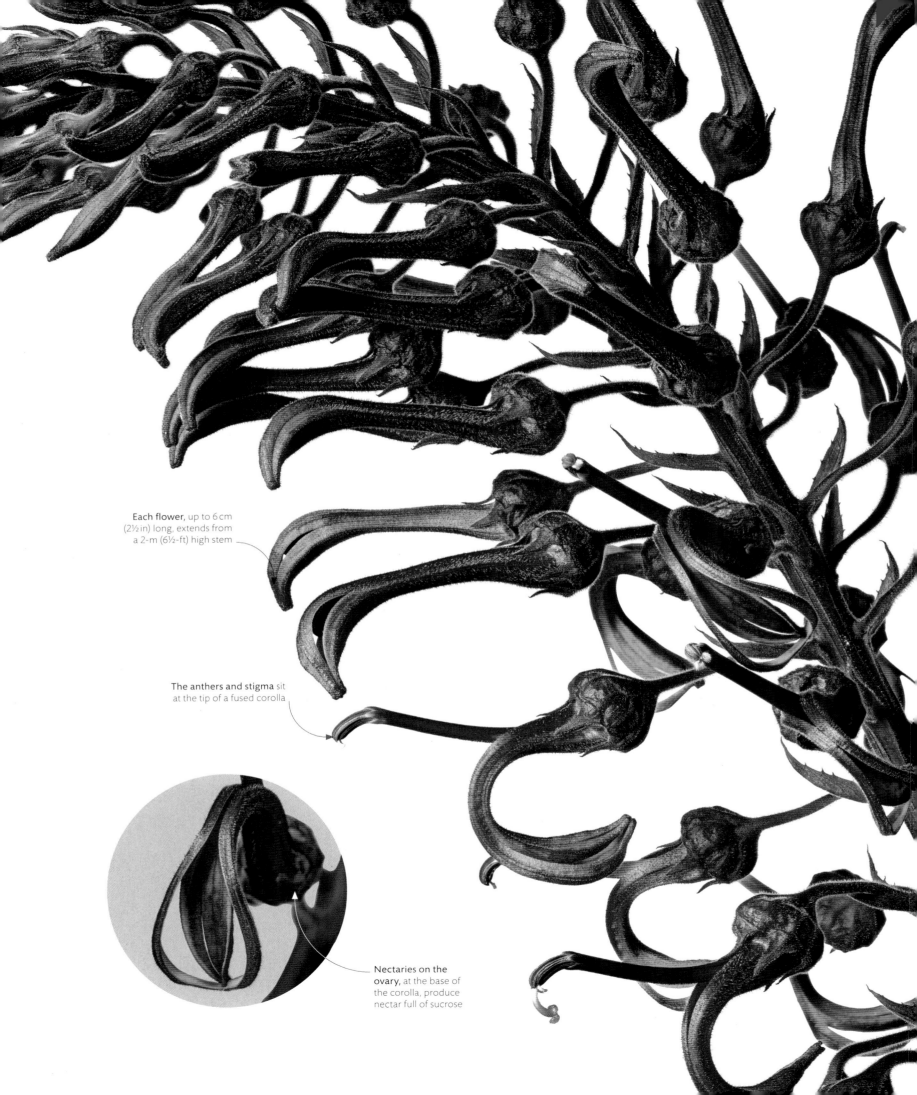

Each flower, up to 6 cm (2½ in) long, extends from a 2-m (6½-ft) high stem

The anthers and stigma sit at the tip of a fused corolla

Nectaries on the ovary, at the base of the corolla, produce nectar full of sucrose

flowers for birds

Many flowers have evolved to attract bird pollinators. They share many common features, including a lack of fragrance, specific vivid colours, and the amount and type of nectar. Long-billed, long-tongued, hovering species, such as hummingbirds, prefer tubular flowers, while other birds, such as honeyeaters and sunbirds, visit blooms that provide convenient platforms on which to perch.

Hummingbirds insert their bills where the floral tube diverges from the petals

Invitation only

Hummingbirds home in on red flowers, such as Chile's devil's tobacco plant (*Lobelia tupa*), shown left. Multiple inflorescences at the ends of stalks, and nectaries hidden within floral tubes, ensure that only certain birds can reach them – and transfer pollen in the process.

Pollen presenters cover the flower head of *Grevillea*, giving it a "tangled" look

Maximizing transfer

Grevillea flower heads make the most of bird pollinators. While probing for nectar, the bills and heads of honeyeaters are brushed by pollen presenters. These are style tips bearing pollen transferred from a flower's own anthers before opening.

GREVILLEA 'COASTAL SUNSET'

PRIMATE POLLINATORS

The traveller's palm (*Ravenala madagascariensis*) seems to be adapted to large mammal pollinators. Several lemur species transfer pollen among plants as they prise apart the tough, protective leaves to take nectar by "paw-dipping", or by drinking directly from the flowers.

RED RUFFED AND RING-TAILED LEMURS

Each individual flower offers nectar to pollinators, transferring pollen grains to fur as the animals feed

The flower spike, 3–13 cm (1–5 in) long, supports hundreds to thousands of flowers that open in sequence

Sturdy flowers

Silver banksia (*Banksia marginata*) flowers develop on robust spike inflorescences that become repositories for the plant's woody seeds. With a mass of nectar-filled flowers opening in sequence, the spikes tempt birds by day and small mammals at night. Pollen transfers to fur just as it does to feathers, and the spikes support the weight of tiny nocturnal creatures as they search for food.

flowers for animals

Birds and insects pollinate many plant species, but mammals also play a crucial role in pollination. Many of these furry pollinators are small, nocturnal creatures such as mice, rats, and shrew-like sengis, lured by the sweet, energy-rich nectar and able to crawl over flowers without damaging them. Even larger carnivores, such as Cape grey mongooses, have been seen raiding flowers – and pollinating them in the process.

Underbelly fur
picks up pollen as the
animal crawls over a
flower spike looking
for nectar

The tiny possums
weigh less than a
hen's egg and are just
9 cm (3½ in) long

Pygmy possums
Prolific nectar and pollen feeders, Australia's pygmy
possums help to maintain many habitats. These tiny
marsupials pollinate banksias, eucalypts, and
bottlebrush plants.

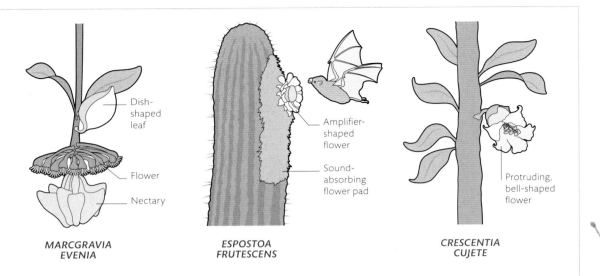

ADAPTED FOR SONAR

Nectar-feeding flying foxes find flowers by scent and sight, but New World leaf-nosed bats rely on echolocation, emitting ultrasonic calls that help them to identify the blooms. The plants that they pollinate have evolved features, such as cushion- or bell-shaped flowers, to bounce the bats' calls back to them more efficiently and let them know where they are. The Cuban rainforest plant *Marcgravia evenia* uses a "satellite dish" leaf to guide bats to its flowers, and Ecuador's *Espostoa frutescens* boosts the audibility of its amplifier-shaped flowers with a "dampener" of fuzzy flower heads.

Dish-shaped leaf

Flower

Nectary

MARCGRAVIA EVENIA

Amplifier-shaped flower

Sound-absorbing flower pad

ESPOSTOA FRUTESCENS

Protruding, bell-shaped flower

CRESCENTIA CUJETE

flying visitors

Birds pollinate many types of plant, but from 500 to 1,000 species of flowering plants (angiosperms) rely on bats to do the same job, particularly in tropical ecosystems and deserts. At least 48 bat and flying fox species around the world have evolved along with the flowering plants on which they depend for food – so much so, that the forms of each have adapted to the other.

The **strong, thick stems** of mangrove flowers are easy for flying foxes to cling to

A **flying fox's entire body** is dusted with pollen when feeding, especially its head and face

A **scar** indicates where a leaf stem dropped off prior to blooming

Fruit bats

Unlike smaller leaf-nosed bats that feed on nectar, flying foxes need large, sturdy flowers to land on as they feed. The cave nectar bat (*Eonycteris spelaea*), an Asian species of flying fox, eats nectar and pollen from plants such as mangroves, and important crops such as banana and durian trees. By clambering onto each flower, the bats get covered in pollen, helping to transfer it easily.

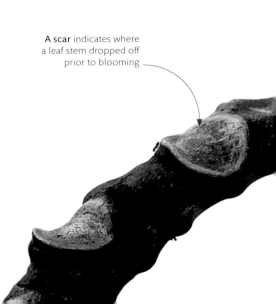

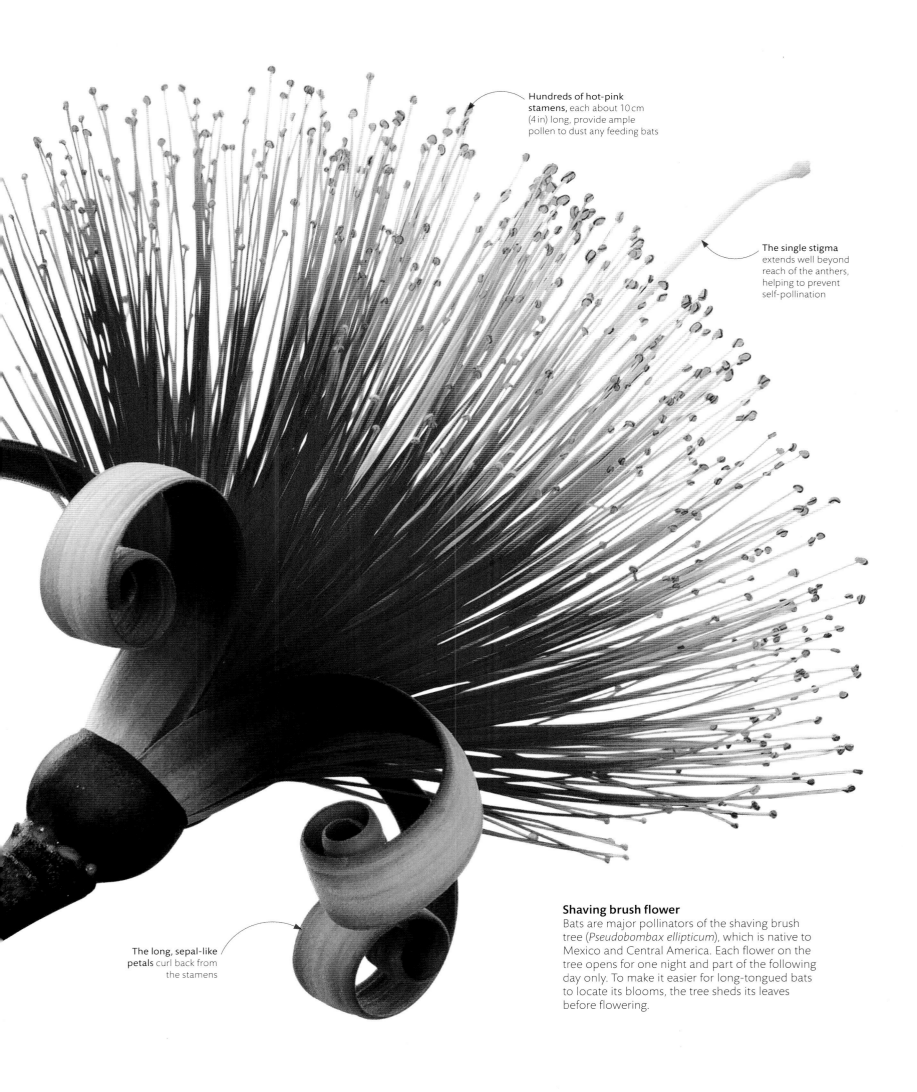

Hundreds of hot-pink stamens, each about 10 cm (4 in) long, provide ample pollen to dust any feeding bats

The single stigma extends well beyond reach of the anthers, helping to prevent self-pollination

The long, sepal-like petals curl back from the stamens

Shaving brush flower

Bats are major pollinators of the shaving brush tree (*Pseudobombax ellipticum*), which is native to Mexico and Central America. Each flower on the tree opens for one night and part of the following day only. To make it easier for long-tongued bats to locate its blooms, the tree sheds its leaves before flowering.

Flower bud
The buds emerge from notches on the edges of the leaf-like stems. Buds usually start to unfold between 10 pm and midnight, in response to the falling night temperature.

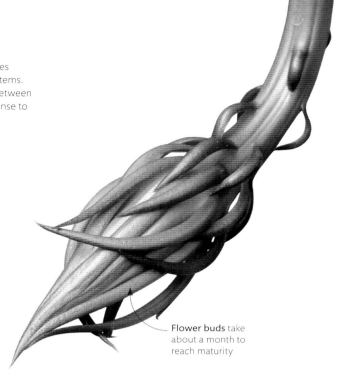

Flower buds take about a month to reach maturity

Epiphyllum oxypetalum

the queen of the night

It seems counter-intuitive to look for cacti in trees, let alone doing so in a wet tropical forest, but that is exactly where one finds the queen of the night. Native to southern Mexico and much of Guatemala, this cactus has stunning blooms with an intoxicating scent that last for one night only.

The queen of the night lives an epiphytic lifestyle, making its home up in the forest canopy. All this species needs for its seeds to germinate is a little humus in a tree hollow or fork between branches. Despite its unusual appearance, the cactus shares many anatomical features with its more cylindrical relatives. What look like long, sprawling leaves are actually the main stems of the cactus. Because it spends its life clinging to the limbs of trees, its stems have evolved a flattened shape to help the cactus grip these precarious surfaces. Its roots not only provide it with water and

nutrients, but also anchor it in place to stop it from tumbling out of the canopy. The queen of the night is spineless. Spines serve to protect cacti from too much sun and the attentions of herbivores, but these are minor problems in the shaded tropical forests where this species grows.

The name queen of the night refers to the plant's massive blossoms, which open only after dark. Extremely fragrant and brilliant white, the flowers are visited by nocturnal sphinx moths. The moths have only one chance to pollinate each flower: by the time the sun rises, the 17 cm (10 in) wide flowers have already wilted. If pollination has been achieved, the flowers are soon followed by bright-pink fruits. Birds and other tree-dwelling animals relish the soft pulp inside the chubby little fruits. Seeds that pass through their digestive tracts are deposited on canopy branches, and so a new generation begins.

Fragrant bloom
The flower's funnel-shaped centre holds a mass of pollen-laden stamens and a long, white stigma. The chemical that gives the flowers their wonderful aroma, benzyl salicylate, is commonly used as a fragrant additive to perfumes.

colour attraction

The scent, size, and shape of flowers all play vital roles in appealing to pollinators, but colour is undoubtedly one of the most important ways a plant has of attracting attention. The colour preferences of insects and birds may sometimes seem strange to us – until we realize that their eyes, being structurally different, are able to discern a different colour spectrum to ours. Bees and many other insects are able to perceive ultraviolet light, which helps them more easily identify the paths to nectar.

White flowers attract nocturnal moths and beetles, also butterflies and flies

Red and orange flowers are favoured by birds

Pink flowers are preferred by butterflies and some moths

Yellow appeals to butterflies, bees, hoverflies, and wasps

Customized colours
Plants have evolved a rainbow of shades to match the visual preferences of their pollinators. While many colours are seen by both insects and birds, not all perceive them the same way – thus bees gravitate to purple while some bird species are drawn to more vivid oranges and reds.

GREEN FLOWERS

Scientists believe that flowers evolved earlier than many insect pollinators, and that their blooms were coloured green, like their surrounding foliage. As relationships developed between plant and pollinator, and competition for pollinators intensified, plants began to adapt floral colours to attract certain species.

GREEN HELLEBORE
(*HELLEBORUS VIRIDIS*)

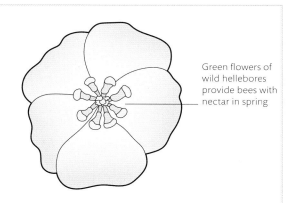

Green flowers of wild hellebores provide bees with nectar in spring

Bluish-purple flowers attract some bees and butterfly species

Purple flowers are most preferred by bees

Blue flowers of all shades are easily seen by bees

Dark purple-brown blooms appeal to wasps

nectar guides

The human eye perceives reflected light as various colours, but many pollinators see the world quite differently. Bees, in particular, discern a specialized range of wavelengths, including the ultraviolet spectrum. This enables them to see features of a flower, such as lines, dots, and other patterns – invisible to the human eye – that guide them directly to the nectar. These "nectar guides" are all-important, not just for the bees but also for the plants, as the bees help the flower to disperse its pollen.

The smallest dots are furthest away from the nectary

Purple dots, along with violet and blue ones, are the three colours most likely to attract bees

The light background contrasts with the dark nectar guides

Large dots tell bumblebees that they are getting closer to the nectary

BEE'S EYE VIEW

Bees lack the photoreceptors necessary to see colours like red, but their ability to discern ultraviolet light transforms seemingly plain surfaces into complex patterns. What we see as a plain yellow marsh marigold (*Caltha palustris*), a bee perceives as a light-coloured bloom with a contrasting dark centre – effectively a "bull's-eye" proclaiming "Land here".

Human's view

Bees' view

IN DAYLIGHT   **IN ULTRAVIOLET LIGHT**

Follow the dots

The extraordinary mottling of the toad lily (*Tricyrtis hirta*) advertises its presence to nectar-seeking insects. Although the dots are also visible to humans in this instance, the increasing size of the dots is especially striking to bumblebees, the toad lily's main pollinators, because bumblebees prefer dots to linear nectar guides.

Immature
green bud

The **nectary** is deep in
the flower, forcing bees
to pass its reproductive
parts to gain access

Glandular hairs
(trichomes) on the
stem deter unwanted
visitors from accessing
the nectaries

Alcea sp.

the hollyhock

With their tall racemes of big, showy flowers, hollyhocks are greatly valued as garden plants. But like the blooms of many flowering plants, hollyhock blossoms bear markings that are hidden from human eyes and are only visible to pollinators able to see the ultraviolet (UV) part of the spectrum.

The 60 or so hollyhock species (*Alcea* sp.) belong to the mallow family (Malvaceae), making them distant cousins of plants such as hibiscus. In summer, hollyhocks produce large, funnel-shaped flowers. The visual appeal of these soaring blooms to the gardener is very different from what attracts pollinators able to detect UV light. Where human eyes may see a plain blossom, the UV-sensitive eyes of bees, the hollyhock's main pollinators, see bull's-eye markings around the centre of the flower. The bull's-eye pattern is produced by special pigments that either

reflect or absorb UV light. As well as bees, many insects, including butterflies, and even some birds and bats, can perceive UV light.

These markings, called nectar guides, are not unique to hollyhocks – many flowers have patterns that are only visible in UV. The markings vary widely, but they all serve the same function: like runway lights, they direct pollinators to a flower's stores of nectar and pollen. Both plant and pollinator benefit from this, as it means that insects spend less time looking for pollen and nectar, and flowers are pollinated more quickly. In fact, insects avoid mutant flowers lacking nectar guides. Some insects that hunt on flowers, such as crab spiders and orchid mantises, resemble nectar guides when seen in UV light. This clever disguise may help to lure insect prey towards the nectar – and their doom.

Fluorescing hollyhock
When a hollyhock is made to fluoresce under UV light, its bull's-eye pattern is revealed. As well as guiding pollinators to nectar, such markings may also help pollinators distinguish between flowers that look much the same to human eyes.

The funnel-shaped flowers, around 10 cm (4 in) across, may be white, pink, red, purple, or yellow

Hollyhock buds flower sequentially rather than all at once, helping to avoid self-pollination

Common hollyhock
Alcea rosea, the common hollyhock, can reach 2.5 m (8 ft) in height, and bears saucer-sized flowers along its upright stem. Native to China, it is now widely cultivated for its attractive blooms.

Emerging flowers are pale pink, signalling high acid and high nectar levels

Older flowers turn blue-purple, meaning less nectar and lower acidity

Dark-pink buds have the highest acidity levels

Floral acid test

The flowers of lungwort (*Pulmonaria officinalis*) emerge pink and turn blue-purple as they age. This colour change is driven by acidity levels within the lungwort flower, which affect the coloured pigments (anthocyanins). The flower's pH changes as it matures, so young, pink, nectar-rich flowers are more acidic than blue-purple ones.

The reddish colour fades on the petals of mature or pollinated blooms

A pinkish-red colour indicates unopen buds or immature flowers with little or no available rewards

colour signals

While certain shades of flowers are known to be more or less attractive to different pollinator species, many plants take the use of colour a step further. By varying the hues of individual flowers at a particular stage and age of development, plants such as honeysuckle (*Lonicera periclymenum*) not only attract the right pollinators, but can direct them towards maturing blooms that contain the most nectar or pollen rewards. In return, the plant receives more visits from passing insects, which results in a much higher proportion of fertilized flowers.

SUBTLE SIGNS

Some flowers deploy subtle colour changes to indicate their status. On spring snowflakes (*Leucojum vernum*), tiny green spots change from green to yellow as the flowers mature. Thought to indicate whether or not the blooms have been pollinated, the spots on these early spring flowers attract bees to a much-needed food source.

Spots fade from green to yellow

Spots are all yellow after pollination

Flowers open with deep green spots

LEUCOJUM VERNUM

Communicating with colour
A combination of scent and colour indicates which honeysuckle flowers are worth visiting. Immature buds blush pinkish-red, while white blooms offer the most pollen and turn yellow after pollination has occurred, when they still provide nectar for bees.

A dark-pink to red colour warns potential pollinators to avoid inner, unopened buds

White flowers are strongly scented to attract night-flying moth pollinators

Yellow, nectar-filled flowers attract long-tongued bees – which may also pollinate any neighbouring white flowers

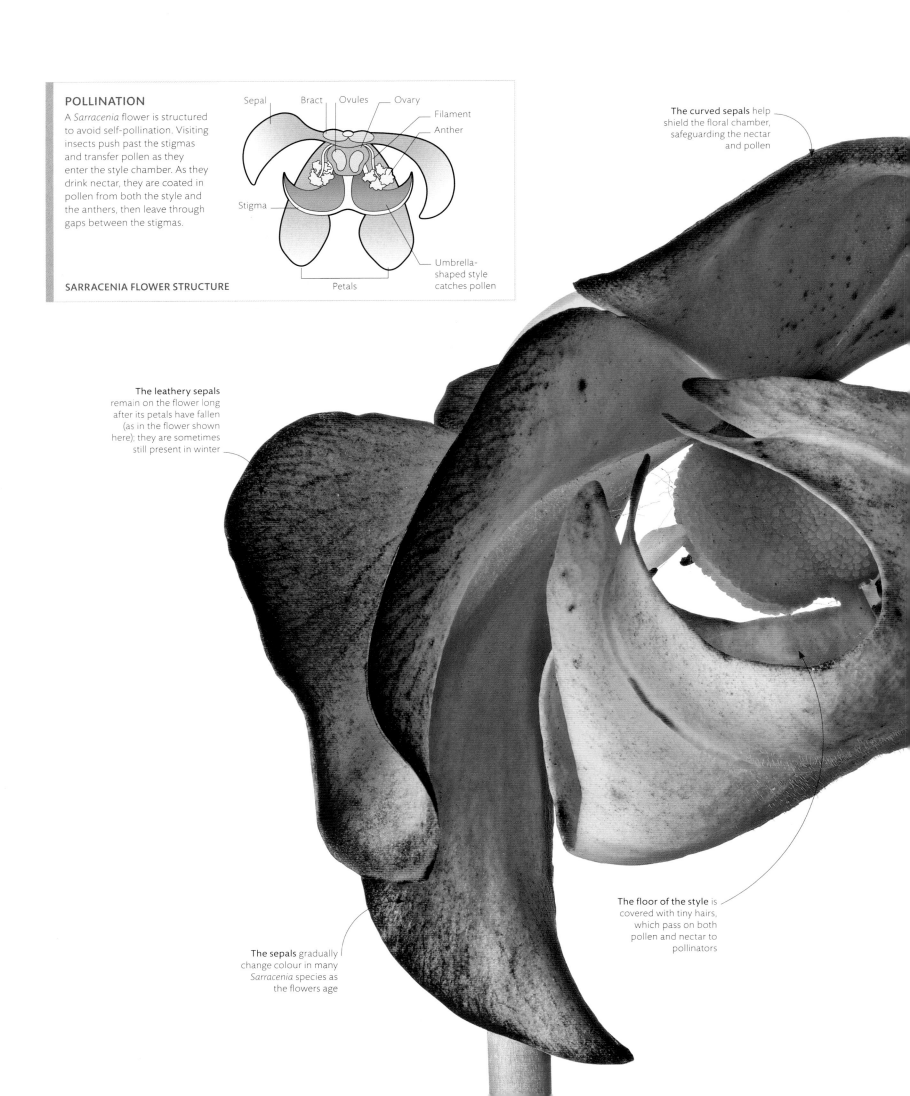

POLLINATION

A *Sarracenia* flower is structured to avoid self-pollination. Visiting insects push past the stigmas and transfer pollen as they enter the style chamber. As they drink nectar, they are coated in pollen from both the style and the anthers, then leave through gaps between the stigmas.

SARRACENIA FLOWER STRUCTURE

Sepal   Bract   Ovules   Ovary

Filament

Anther

Stigma

Umbrella-shaped style catches pollen

Petals

The curved sepals help shield the floral chamber, safeguarding the nectar and pollen

The leathery sepals remain on the flower long after its petals have fallen (as in the flower shown here); they are sometimes still present in winter

The sepals gradually change colour in many *Sarracenia* species as the flowers age

The floor of the style is covered with tiny hairs, which pass on both pollen and nectar to pollinators

Unique style

Single *Sarracenia* flowers develop on separate stems from the plant's carnivorous pitchers, which are often closer to the ground. To reduce the risk of ingesting a valuable pollinator, the flowers bloom in spring, well before the pitchers become active in summer. The style may be a strange shape to prevent self-pollination, but it does allow the plant to hybridize (cross-pollinate) easily.

The flower hangs upside down, due to the weight of the umbrella-shaped style

The style curls around the developing ovary

NEW WORLD PITCHER PLANTS

Pitcher plants belong to the Sarraceniaceae family. This group of three genera – *Darlingtonia*, *Heliamphora*, and *Sarracenia* – contains 34 species, and many of them are highly endangered. All grow in boggy areas with poor soil – which is why they need nutrients from the insects that they entrap.

SARRACENIA DRUMMONDII

restricted entry

Carnivorous pitcher plants attract insects to ingest, but they also need pollinators to come and go in order to reproduce. To do this, the pitcher plant's flowers are physically separated from its deadly traps, not only by space but also by time, because they bloom before the traps become active. Their unique structure also controls the way in which pollinators enter and exit the flowers.

fragrant traps

Many plants use floral fragrance to attract pollinators. Some, however, go a step further and emit irresistible odours that lure insects to flowers in order to trap them inside for "forced pollination". Hundreds of orchid species – including nearly 300 greenhood orchids (*Pterostylis* spp.) – employ this method to ensure cross-pollination, and access a wider gene pool.

A sepal and two petals are fused internally on the hood-like galea that covers the sexual organs

A hinged labellum, or lip, traps the insect when it moves towards the lure at the base

Two fused sepals form the front of the orchid, ending in elongated "points" either side of the galea

Translucent stripes on the galea allow in filtered light, guiding insects to the back of the flower

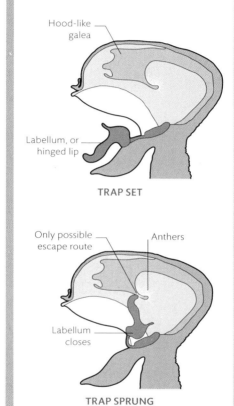

TRAPPING MECHANISM

When a gnat begins crawling along the labellum of a greenhood orchid, the lip flexes and tips the insect inside. This traps it within the gynostemium – a reproductive structure found in several plant families that consists of fused stamens and pistils. Once in this column, the insect can only exit by squeezing past the anthers, which press a mass of pollen, known as a pollinium, onto its back. The insect carries this to the next orchid, pollinating that flower in the process.

Hood-like galea

Labellum, or hinged lip

TRAP SET

Only possible escape route

Anthers

Labellum closes

TRAP SPRUNG

Chemical attraction

Greenhood orchids (*Pterostylis* spp.) are native to parts of Southeast Asia, Australia, and New Zealand, but *P. tenuicauda* is only common in New Caledonia. Most of the insect visitors to these flowers are male fungus gnats. It is thought that, to attract them, the scent given off by the blooms mimics the pheromones of tiny female flies.

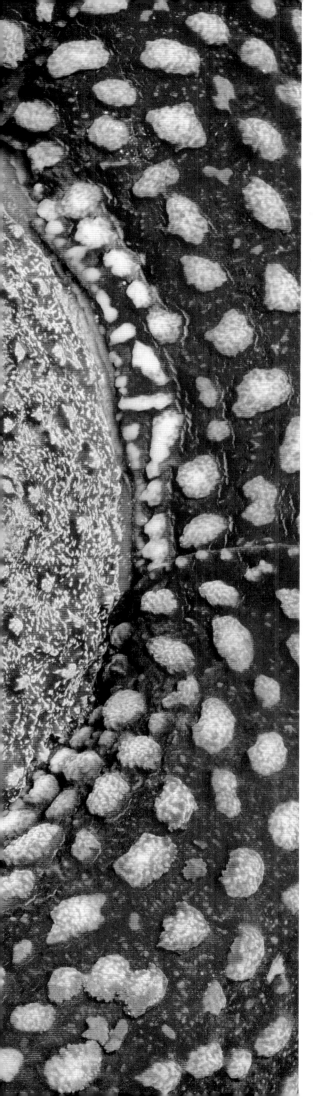

The red coloration and textured surface are thought to mimic the look and feel of decaying flesh

Short-lived grandeur
Corpse flower blooms may have evolved to such a huge size to make them easier for pollinators to find. For all its grandeur, the bloom of the corpse flower is an ephemeral event, lasting barely a week.

Rafflesia arnoldii

the corpse flower

At 1 m (3 ft) across and weighing 11 kg (24 lb), the corpse flower is the world's largest single flower. Despite its size, this rare flower is usually smelled before it is seen. Native to the diverse rainforests of Sumatra and Borneo, the plant mimics rotting meat in both fragrance and appearance.

The vast bloom is all that is visible of the *Rafflesia arnoldii*. This parasitic species has no stem, no leaves, and no roots. Invading the food-conducting vascular tissues of forest vines, it mainly consists of thread-like tissues living in and around its host's cells, from which it obtains the nutrients and water it needs. The corpse flower cannot live without its host, so it rarely has any serious effect on the vine's health.

At flowering time, tiny buds form on the vine stem and gradually swell to look like large purple or brown cabbages. The flower buds take up to a year to develop,

Mystery plant
Within the central cavity of the corpse flower is a disk covered with projections whose function is uncertain. The anthers and stigma lie below this disk. The plant's gloopy pollen dries on the back of the flies and may remain viable for weeks.

during which time they are sensitive to disturbance. Flowers are either male or female, so they need to appear relatively near each other in order to reproduce. The principal pollinators of *Rafflesia arnoldii* are carrion flies. Lured into male flowers by the false promise of rotting meat, the flies become covered in gooey blobs of pollen. When they visit female flowers, they are corralled into a narrow crevice that forces them to brush against the stigma and transfer the sticky pollen. The rarity of these plants means that the odds of a male and female plant being in flower at the same time and within flying distance of one another are slim; sexual reproduction does not happen often.

This species relies on intact forest for survival. Deforestation may be pushing the species to the brink of extinction, but its rarity and cryptic lifestyle make it hard to properly assess numbers in the wild.

special
relationships

Mutualism – an association in which two different organisms each benefit from the actions of the other – is an intrinsic part of the botanical world, from fungi that nourish forest root systems to the pollination of flowers by animals. Over time, highly specialized relationships have evolved, resulting in structural changes on the part of flowering plants, and changes in the behaviour of the animals that depend on them.

The blue inner petal segment contains the anthers and style in a dart-shaped structure

Each flower has three orange sepals that stand up like a bird's crest

Bird of paradise
The South African *Strelitzia reginae* has evolved an inflorescence that resembles an exotic bird's head. Commonly known as the bird of paradise plant, its vivid pointed flower parts are specifially adapted to pollination by birds.

The scale-like structure at the base is a third petal, concealing the nectary

The orange sepal bends back away from the blue petals

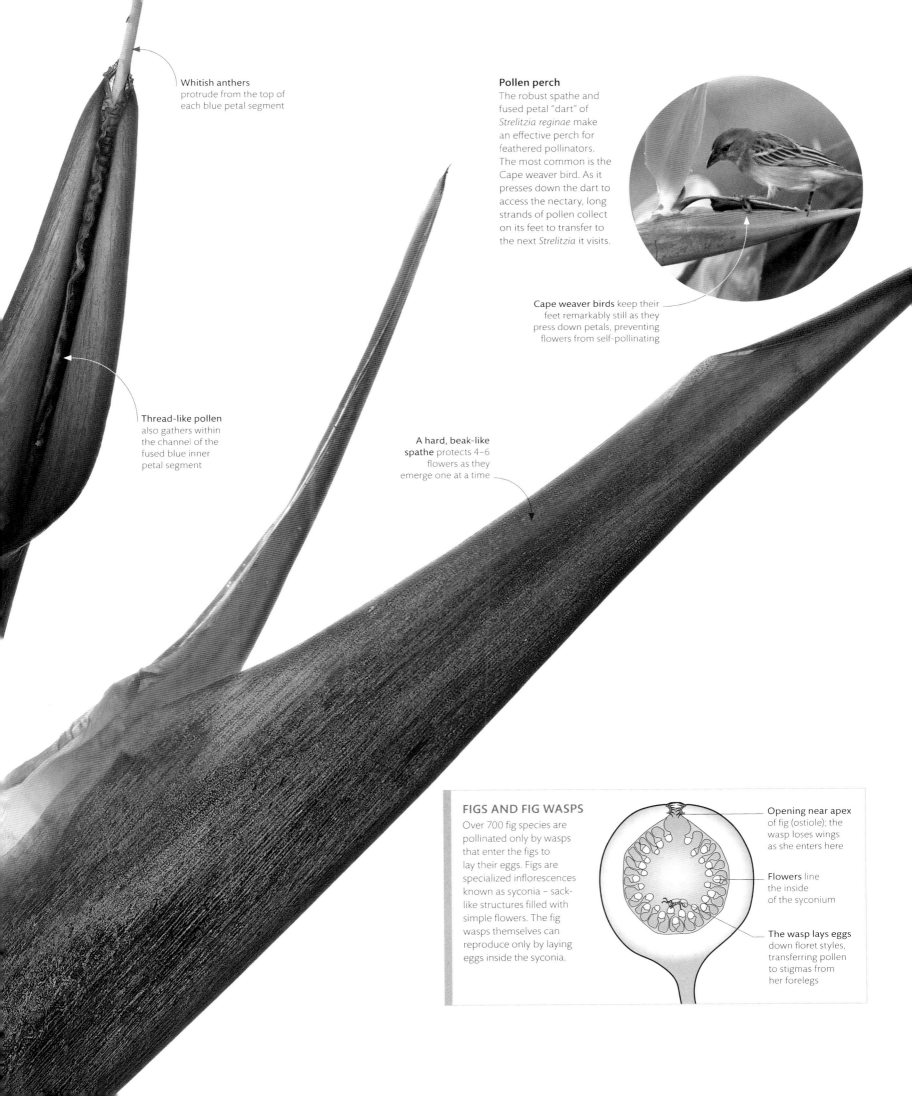

Whitish anthers protrude from the top of each blue petal segment

Thread-like pollen also gathers within the channel of the fused blue inner petal segment

A hard, beak-like **spathe** protects 4–6 flowers as they emerge one at a time

Pollen perch

The robust spathe and fused petal "dart" of *Strelitzia reginae* make an effective perch for feathered pollinators. The most common is the Cape weaver bird. As it presses down the dart to access the nectary, long strands of pollen collect on its feet to transfer to the next *Strelitzia* it visits.

Cape weaver birds keep their feet remarkably still as they press down petals, preventing flowers from self-pollinating

FIGS AND FIG WASPS

Over 700 fig species are pollinated only by wasps that enter the figs to lay their eggs. Figs are specialized inflorescences known as syconia – sack-like structures filled with simple flowers. The fig wasps themselves can reproduce only by laying eggs inside the syconia.

Opening near apex of fig (ostiole); the wasp loses wings as she enters here

Flowers line the inside of the syconium

The wasp lays eggs down floret styles, transferring pollen to stigmas from her forelegs

Dark sepals are thought to imitate an insect's "flight pattern", causing male bees to charge

White-tipped sepals attract male *Centris* bees, who respond aggressively to moving white objects

Attracting both sexes

Some *Oncidium* orchids assume a colour, called bee-UV-green, and shape similar to that of butterfly vine flowers, which attract female *Centris* bees in search of oil and pollen. When fluttering in the wind, however, some parts look like enemy insects. As male bees attack them, they get covered in pollen.

The petals of this *Oncidium* are bee-UV-green – a colour visible only to bees, who mistake it for a butterfly vine

The orchid lip, or labellum, evolved to resemble paddle-shaped butterfly vine petals

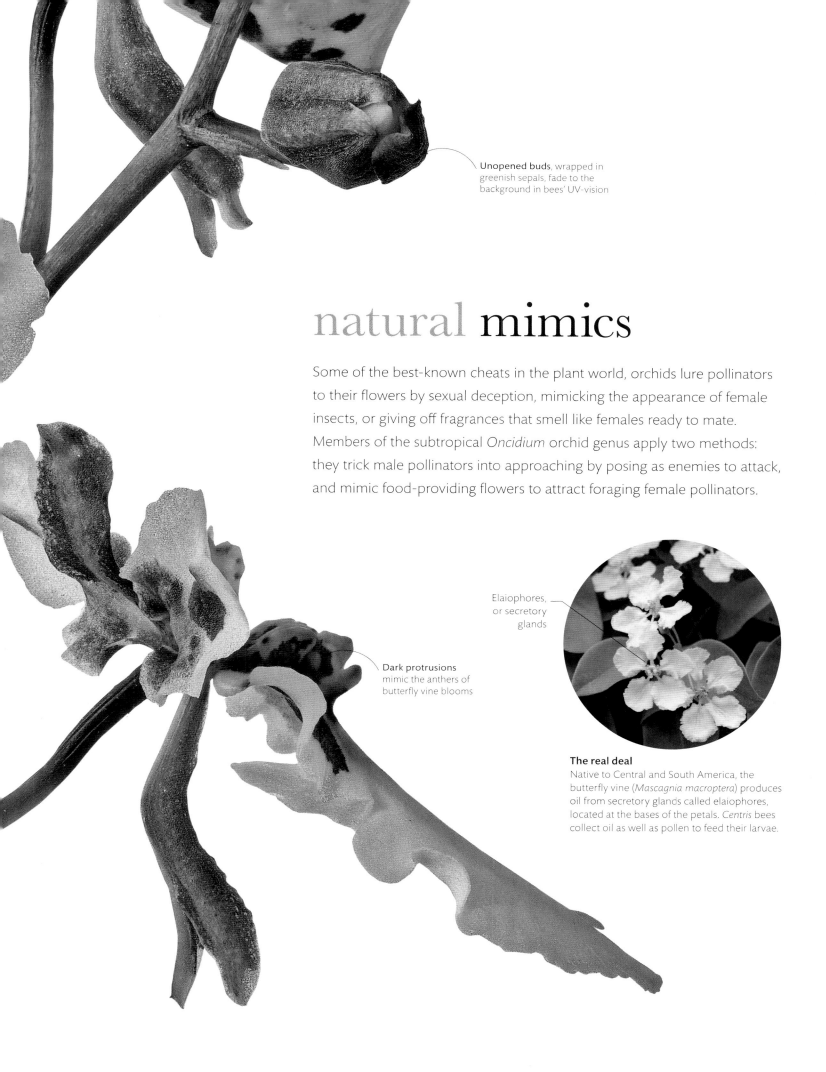

Unopened buds, wrapped in greenish sepals, fade to the background in bees' UV-vision

natural mimics

Some of the best-known cheats in the plant world, orchids lure pollinators to their flowers by sexual deception, mimicking the appearance of female insects, or giving off fragrances that smell like females ready to mate. Members of the subtropical *Oncidium* orchid genus apply two methods: they trick male pollinators into approaching by posing as enemies to attack, and mimic food-providing flowers to attract foraging female pollinators.

Elaiophores, or secretory glands

Dark protrusions mimic the anthers of butterfly vine blooms

The real deal
Native to Central and South America, the butterfly vine (*Mascagnia macroptera*) produces oil from secretory glands called elaiophores, located at the bases of the petals. *Centris* bees collect oil as well as pollen to feed their larvae.

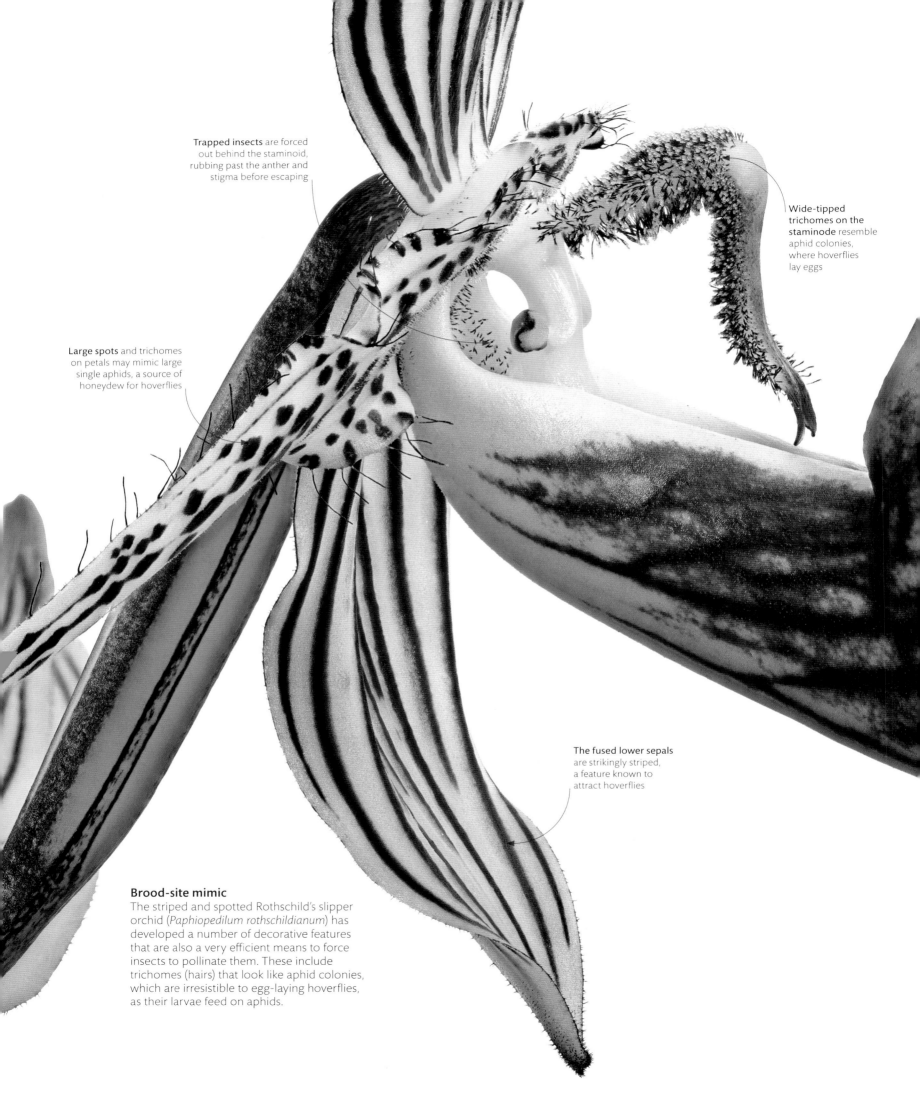

Trapped insects are forced out behind the staminoid, rubbing past the anther and stigma before escaping

Wide-tipped trichomes on the staminode resemble aphid colonies, where hoverflies lay eggs

Large spots and trichomes on petals may mimic large single aphids, a source of honeydew for hoverflies

The fused lower sepals are strikingly striped, a feature known to attract hoverflies

Brood-site mimic
The striped and spotted Rothschild's slipper orchid (*Paphiopedilum rothschildianum*) has developed a number of decorative features that are also a very efficient means to force insects to pollinate them. These include trichomes (hairs) that look like aphid colonies, which are irresistible to egg-laying hoverflies, as their larvae feed on aphids.

Following fashion
These two *Paphiopedilum* hybrids sport features similar to that of the Rothschild's slipper orchid, attracting pollinators with enticing stripes, and spots that can be mistaken for aphids.

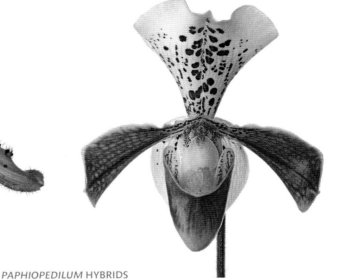

PAPHIOPEDILUM HYBRIDS

After landing, hoverflies often fall into the labellum, or lip, and become trapped

designed to deceive

Some plants reward pollinators, but others operate through deception, fooling pollinators with the false promise of a sweet reward. In the case of slipper orchids, lures vary from spots or hairs that mimic aphids that appeal to predators, to holes that resemble tunnels, which attract bees looking for nest sites. Once finding themselves inside a cavity, visiting insects must locate the one and only way out, which forces them past the plant's sex organs, making them pollinate flowers for no reward.

The pouch-shaped labellum – actually a third petal – resembles a slipper

Elongated, spotted petals set almost horizontally, increase the plant's "advertising space" to attract pollinators

CLOSED FLOWER STRUCTURE

The corollas of *Lamprocapnos* flowers are much smaller than those of many of its relatives in the poppy family – they are also inconspicuous, and may be colourless. They completely enclose the anthers and stigmas, which are so tightly packed together that they may touch. This makes it easy for pollen to be transferred within the flower, allowing self-pollinated seeds to develop.

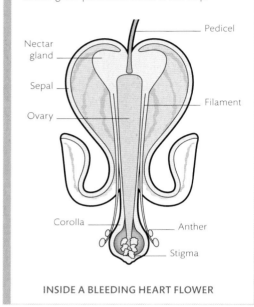

Nectar gland

Sepal

Ovary

Pedicel

Filament

Corolla

Anther

Stigma

INSIDE A BLEEDING HEART FLOWER

The terminal bud of the raceme's main stem (peduncle) keeps growing

The tip of the corolla surrounds the anthers and stigma

Heart-shaped blooms

The flowers of bleeding heart (*Lamprocapnos spectabilis*) may be pollinated with or without the help of insects. The flowers produce nectar, attracting bumblebees, but the close proximity of the male and female parts inside the tubular corolla makes self-pollination possible when pollinators are in short supply.

self-pollinating flowers

It takes a lot of energy for a plant to produce the colour and nectar necessary to attract pollinators, and sometimes it is advantageous for a flower to fertilize itself. Self-pollination allows plants that grow in challenging conditions to preserve the traits that help them to succeed in those environments. It may also be the reason why small or sparse populations have survived. For many species, self-fertilization is a useful "backup plan", to be deployed at times when pollinators are scarce.

A single raceme of a bleeding heart plant may produce 3–15 pendulous flowers

The pink sepals flex back as the flower matures

Anthers are very close to the stigma

The petals of open dog violet flowers are large and attract pollinators

Fallback strategy
Some plants have two types of flower. If the spring blooms of the dog violet (*Viola riviniana*) are not pollinated, all is not lost. In autumn, it produces more flowers at soil level. These closed (cleistogamous) blooms self-fertilize and produce seeds without the wind or insect activity.

Rosa Centifolia.   *Rosier à cent feuilles.*

P. J. Redouté   Langlois.

plants in art

blooms for royalty

In the golden age of botanical illustration spanning the late 18th and early 19th centuries, leading artists were favoured in the royal courts of Europe and achieved international status for their work. Their watercolours could be reproduced with exquisite accuracy thanks to major advances in printing and copperplate engraving techniques.

Often referred to as the "Raphael of flowers", Belgian artist Pierre-Joseph Redouté published more than 2,000 plates depicting 1,800 plant species during his lifetime. He studied plant anatomy with French aristocrat Charles Louis L'Héritier de Brutelle, and floral art under Gerard van Spaendonck, who was miniature painter to Louis XVI of France. Redouté was first appointed as court artist and tutor to Queen Marie Antoinette. After the French Revolution he was engaged in Empress Josephine's project to create one of the finest gardens in Europe at the Château de Malmaison. Her rose garden included 200 different types of rose, many of which appear in Redouté's three volumes of *Les Roses* – a work that is still used today to identify older varieties.

In 1790, the Austrian artist Franz Bauer was appointed as the first botanical artist in residence at the Royal Gardens at Kew. Bauer was both a scientist and a skilled artist, whose work includes microscopic studies of plant anatomy.

Jacinthe Double, 1800
The work of Flemish artist Gerard van Spaendonck combines traditional Dutch mastery of flower painting with French sophistication. *Jacinthe Double* (left) is one of 24 plates using stipple engraving that was developed further by Redouté.

Rosa centifolia, c.1824
A hybrid rose with a clear, sweet fragrance, "the rose with a hundred petals" is featured in a stipple engraving from Pierre-Joseph Redouté's *Les Roses* (left). Redouté perfected the technique of stipple engraving copper plate with fine dots to reproduce the gradations of colour seen in his flawless watercolours and allow the "light" of the paper to shine through. The prints were then finished by hand in watercolour.

❝ …among the flowers which have received, in the highest degree, the gift of mutability, none can be compared to the Rose… ❞

CLAUDE ANTOINE THORY, FOREWORD, *LES ROSES*, 1817

TEMPERATURE CHANGES

Many flowers open and close because the fluids in their cells respond to temperature changes, which make the cells expand and contract, and expanding cells create surface pressure that forces petals open. The inner and outer sides of tulip petals differ in temperature by up to 10°C (50°F). As sunlight warms the flower, the surface temperature of the inner petal rises and its cells expand, pushing the flower open. As the temperature drops, cells on the inner surface contract first, pulling the flower closed again.

Petal

LOW TEMPERATURE

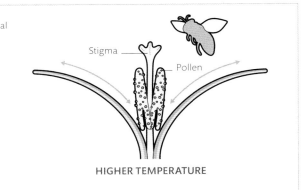

Stigma

Pollen

HIGHER TEMPERATURE

Petals open and close in response to light and heat

Cone-shaped receptacle containing the stigmas

Copious anthers produce around a million pollen grains per bloom

Each lotus flower has 18–28 petals

Opening hours

A lotus (*Nelumbo* sp.) flower lasts for 3–4 days, opening at dawn and closing at dusk. The first day it opens only partially to enable its stigmas to receive pollen, then it closes completely. The next two mornings it opens more fully, releasing scent to attract bees, flies, and beetles.

closing at night

Some flowers open or close in response to external stimuli, such as touch, or changes in light, temperature, or humidity. These factors trigger physical reactions in many species, but flowers that close at night may also have a reproductive agenda. By closing, they protect their pollen and sex organs from the elements, and reduce the risk of being eaten or damaged by nocturnal predators. They therefore increase the chances of attracting pollinators by day.

The petals close rapidly as the light fades and temperatures drop

The petals close so tightly on the first day that the flower resembles a bud

Each flower has only two sepals

Closing at night
When lotus flowers (*Nelumbo* sp.) close at night, chemical changes inside the receptacles generate heat, creating temperatures up to 40°C (104°F) higher in the flower than the outside air. The heat releases scent, which the lotus, lacking nectar, needs in order to attract pollinators when it opens the next day.

Sepals protect the flower bud as it emerges

Glandular hairs cover the sepals and provide extra protection for the flower bud

The ovary, the swollen area at the base of the pistil, develops into a hip after pollination

bud defences

Buds are usually protected by sepals, but some plants use tiny hairs (trichomes) to strengthen their defences. These are thought to trap a layer of air around the bud, which insulates it from the elements and regulates temperature and moisture content. To further deter pests, some hairs release chemicals when they are touched.

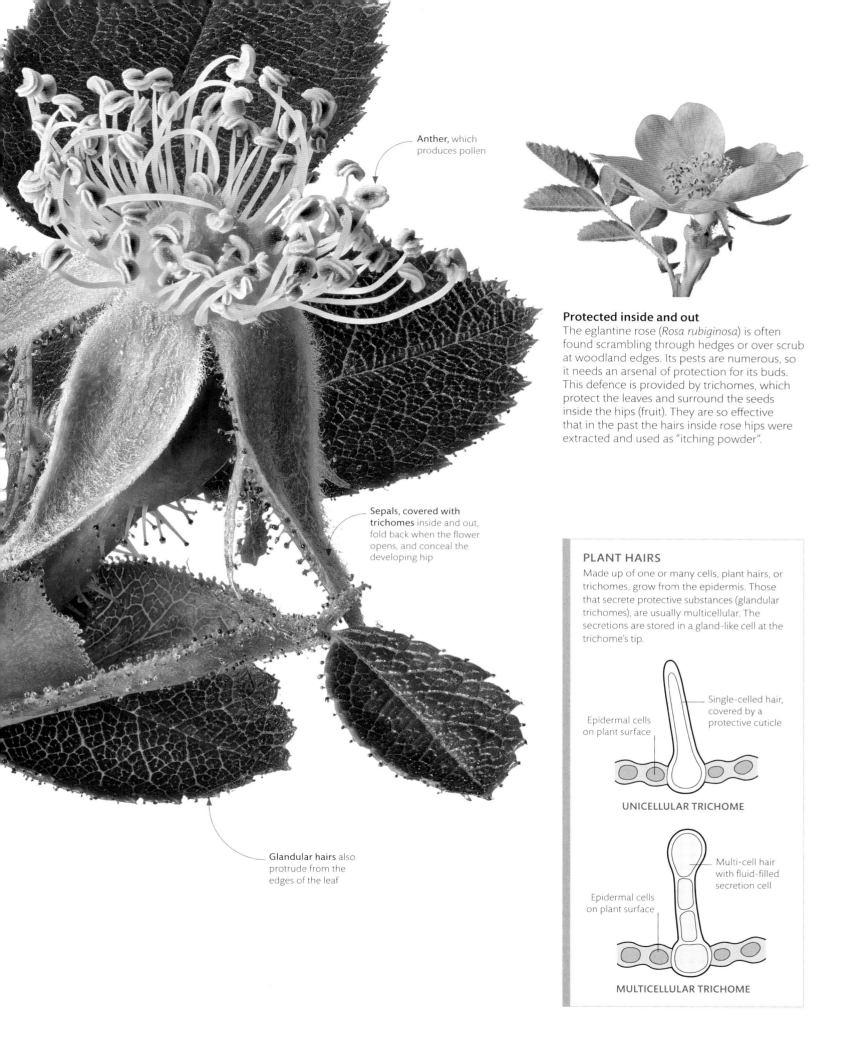

Anther, which produces pollen

Protected inside and out

The eglantine rose (*Rosa rubiginosa*) is often found scrambling through hedges or over scrub at woodland edges. Its pests are numerous, so it needs an arsenal of protection for its buds. This defence is provided by trichomes, which protect the leaves and surround the seeds inside the hips (fruit). They are so effective that in the past the hairs inside rose hips were extracted and used as "itching powder".

Sepals, covered with trichomes inside and out, fold back when the flower opens, and conceal the developing hip

PLANT HAIRS

Made up of one or many cells, plant hairs, or trichomes, grow from the epidermis. Those that secrete protective substances (glandular trichomes), are usually multicellular. The secretions are stored in a gland-like cell at the trichome's tip.

Single-celled hair, covered by a protective cuticle

Epidermal cells on plant surface

UNICELLULAR TRICHOME

Multi-cell hair with fluid-filled secretion cell

Epidermal cells on plant surface

MULTICELLULAR TRICHOME

Glandular hairs also protrude from the edges of the leaf

Flower buds
flush pink as
they develop

Spiky protection
The spines on a teasel
flower are actually very stiff,
sharp bracts that protect
its developing buds.

Anthers and filaments
protrude from pale-pink
to purple tubular corollas

Phased flowering
Protected by its spiny bracts, a
single teasel (*Dipsacus fullonum*)
inflorescence can produce
around 2,000 blooms that
begin to open in a ring around
the middle. Bands of flowers at
the top and bottom mature
later, weeks after the blooms
in the central ring have died.

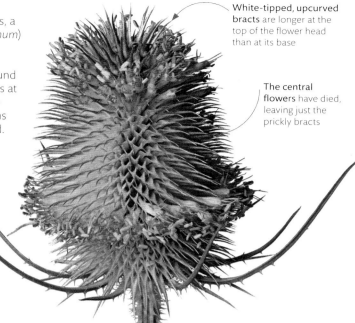

White-tipped, upcurved
bracts are longer at the
top of the flower head
than at its base

The central
flowers have died,
leaving just the
prickly bracts

armoured flowers

When it comes to survival, plants are at a disadvantage – they cannot
move away or hide when threatened by predators. Many plants protect
their leaves and stems with spines, prickles, or thorns, but the common
teasel (*Dipsacus fullonum*) has also developed sharp defences on its flower
heads. This "body armour" allows pollinators to visit the open flowers,
while protecting the buds and developing seeds.

Long, prickly bracts curve
up around the flower head
to form a protective cage

MULTI-PURPOSE SPINES

The spiny bracts of some flowering
plants, such as burdock (*Arctium*
sp.), serve a dual purpose. Not only
do they protect the flower head by
repelling potential predators, but
the hooked tips of the resulting
burrs – the inspiration for the
hook-and-loop fastener – also latch
onto the fur of passing animals. This
helps to disperse the burdock seeds
over a wide area.

Protective
bracts end in
hooked tips

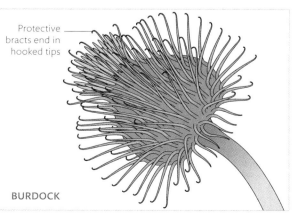

BURDOCK

colourful bracts

Apart from the spectacular autumn displays shown by the leaves of certain trees, colour in the plant world is most strongly associated with flowers. Modified protective leaves called bracts can be just as vibrant as flowers, though, and are often mistaken for flower parts, particularly in hot-climate species. Bracts can also function as brightly hued petals, such as in crimson Central American poinsettias, and their colours prove irresistible to pollinators.

Mistaken identity

The bracts of many tropical species often overshadow the nondescript flowers they protect. The South American hanging lobster claw plant (*Heliconia rostrata*) features strikingly bright crimson and yellow bracts, which attract hummingbirds to pollinate the tiny flowers they enclose.

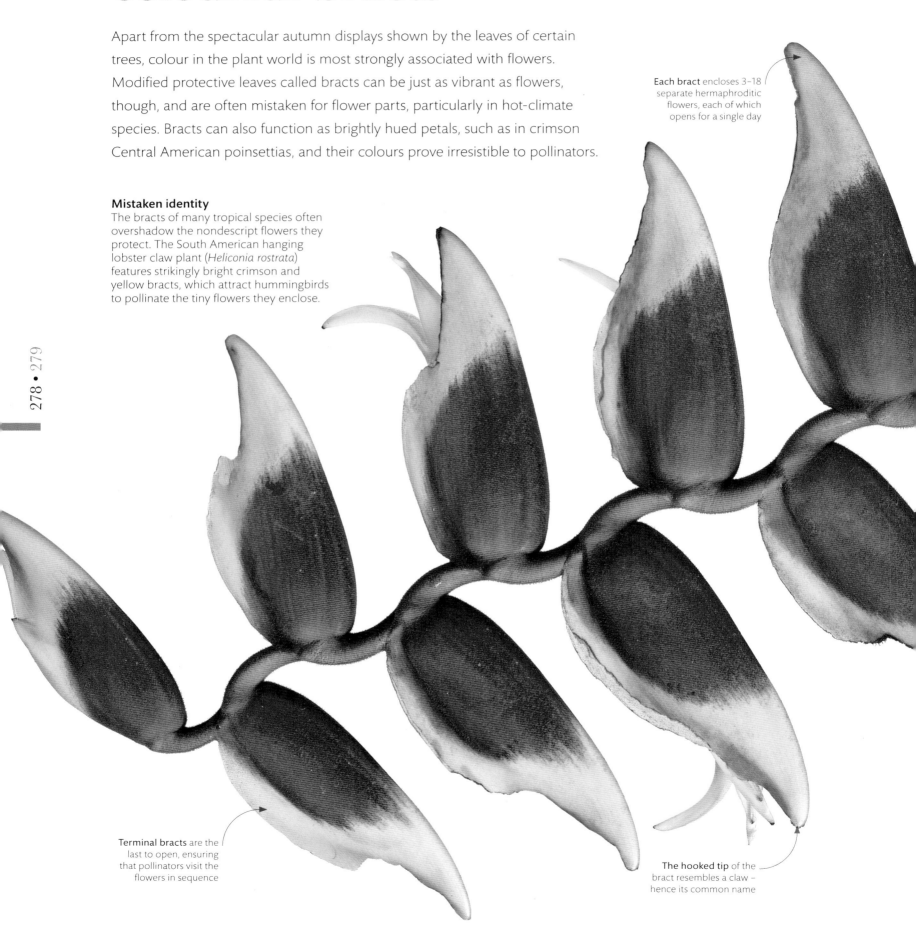

Each bract encloses 3–18 separate hermaphroditic flowers, each of which opens for a single day

Terminal bracts are the last to open, ensuring that pollinators visit the flowers in sequence

The hooked tip of the bract resembles a claw – hence its common name

Crimson-coloured stems add to the vivid inflorescence display

The dominant red colour on the top of the bracts attracts pollinators from above

Pouch-like bracts conceal delicate purple flowers with yellow spots; the colours of the bracts and flowers attract insects

A single sepal protrudes from each flower, allowing access to the nectar

Red, upturned bracts conceal tubular flowers that appeal to hummingbirds

Lighting up the shade
Tropical species use shape as well as vibrant bracts to stand out in shade. Species such as Peruvian *Ruellia chartacea* and the Malaysian beehive ginger (*Zingiber spectabile*) flower in the understorey, but use very different bract arrangements to attract pollinators.

RUELLIA CHARTACEA

ZINGIBER SPECTABILE

Dutchman's Pipe
Mary Vaux Walcott contributed watercolour illustrations to the book *North American Pitcher Plants*, published by the Smithsonian Institution in the US in 1935. This artwork features a species of *Aristolochia* commonly known as the Dutchman's Pipe. The name is taken from the shape of the flowers that resemble smoking pipes once common in the Netherlands and Northern Germany.

plants in art

american enthusiasts

The expansion of the North American railroads during the 19th century gave adventurers, naturalists, and scientists access to the diverse and unexplored habitats of the vast continent. Avid photographers and artists were drawn to remote areas, such as the Rocky Mountains, to capture images of landscapes and wildlife. Notable among these were intrepid female painters who produced remarkable collections of botanical artworks.

Born into a prosperous Quaker family from Philadelphia, Mary Vaux Walcott (1860–1940) first visited the Canadian Rockies on a family holiday in 1887, and was captivated by the landscape. Subsequently returning during most summer holidays, she revelled in the outdoor life and became a keen mountain climber and amateur naturalist. It was these interests that she would combine with her lifelong passion for painting.

On one visit to the Rockies, Walcott was asked by a botanist to paint a rare blooming plant, and the results led her to pursue botanical illustration. For many years she would traverse the rugged terrains of North America, seeking out significant and new species of wild plants, and creating hundreds of watercolour paintings. Around 400 of these were reproduced in a five-volume book set titled *North American Wildflowers*, published by the Smithsonian Institution between 1925 and 1929. Walcott received much acclaim for the work, and her very engaging and botanically accurate images, and was hailed as the "Audubon of Botany".

On some of her expeditions Walcott was joined by her childhood friend Mary Schäffer Warren (1861–1939), who shared her sense of adventure and talent for painting. The publication *Alpine Flora of the Canadian Rocky Mountains* (1907), inspired by her late naturalist husband, featured many of Warren's striking watercolours of plants and flowers. The works produced by these trailblazing women represented a new age of discovery and revealed to the world the true, and little known, beauty of the flora of North America.

Formal education
Roses on a Wall (1877) is a painting by the popular Philadelphia artist George Cochran Lambdin. Renowned for his formal paintings of flowers, Lambdin is known to have had Mary Schäffer Warren as a pupil. Some historians believe that Mary Vaux Walcott may also have studied with him.

> 66 ... to collect and paint the finest specimens obtainable, and to depict the natural grace and beauty of the plant without conventional design. 99

MARY VAUX WALCOTT

Reproduction without flowers
Male and female cones may be produced by
separate plants, but when both occur on the same
tree, they are often found in different parts of the
canopy to promote cross-pollination. On the Atlas
cedar (*Cedrus atlantica*), male pollen cones appear
mainly on the lower branches. High above them,
the female cones are more likely to receive pollen
blown from a neighbouring tree.

Male cedar cones
grow to about 8 cm
(3 in) long

The soft scales of male
cones release clouds
of pollen in autumn

Pollen grains collect
on the needles
before being blown
to female cones

Seed-bearing cones

The female cones of Atlas cedars (*Cedrus atlantica*) may take up to two years to mature. The process of fertilization itself often takes a year to complete, as male pollen-grain tubes push slowly beneath the female cone's scales to deliver sperm to ovules. In the following months, tiny winged seeds develop on the underside of the scales.

Young, green female cones become a woody barrel shape as the seeds develop on their scales

Each broad scale releases two winged seeds

cone reproduction

Although they produce pollen and ovules, gymnosperms – an ancient group of plants that includes cycads, ginkgos, and conifers – share little in common with flowering species. They reproduce by means of male and female cones, creating seeds over a much longer period. The term "gymnosperm" literally means "naked seed", a term that refers to the ovules of female cones, which emerge fully exposed and are not surrounded by a protective ovary.

MALE AND FEMALE CONES

In most gymnosperms, male and female cone structure is different. A male cone usually lives for a few days. It is softer-textured, longer, and slimmer than the female cone, with bract scales arranged in a spiral around a central stem; each scale contains a pollen sac on its lower surface. Female cones are wider and more substantial, with spirally arranged ovule-bearing scales. Each scale carries one or more ovules, and these develop into seeds once they have been pollinated.

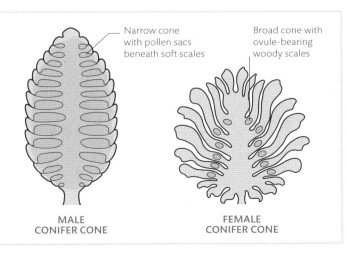

Narrow cone with pollen sacs beneath soft scales

Broad cone with ovule-bearing woody scales

MALE CONIFER CONE

FEMALE CONIFER CONE

seeds. and fruits

seed. a plant's unit of reproduction, from which another such plant can develop.

fruit. the structure that surrounds a plant's seeds, often sweet, fleshy, and edible.

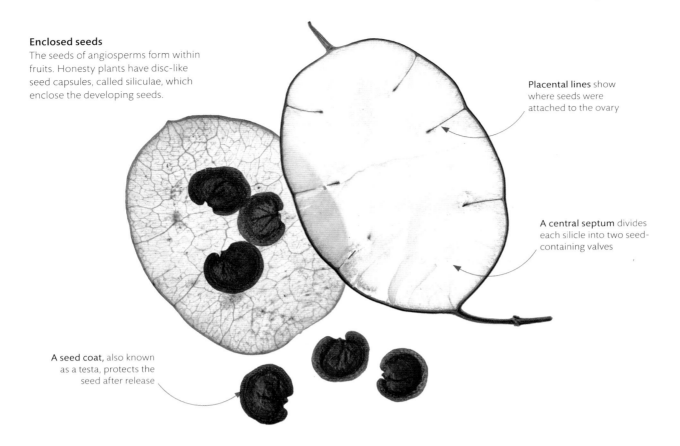

Enclosed seeds
The seeds of angiosperms form within fruits. Honesty plants have disc-like seed capsules, called siliculae, which enclose the developing seeds.

Placental lines show where seeds were attached to the ovary

A central septum divides each silicle into two seed-containing valves

A seed coat, also known as a testa, protects the seed after release

seed structure

Whether they bear cones or fruit, all non-flowering plants (gymnosperms) and flowering ones (angiosperms) reproduce by seed. And while there are differences in how the naked seeds of a conifer develop compared with the enclosed seeds of a flowering plant, the seeds of both have the same basic structure – an outer seed coat, stored nutrients, and a developing embryo.

The silvery septums remain on the plant long after the valves have fallen away

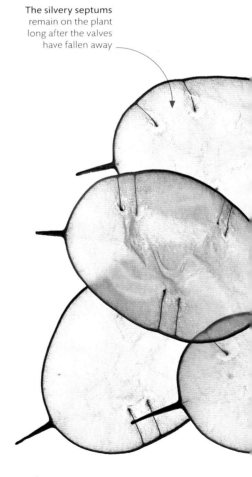

INSIDE SEEDS

All seeds have seed leaves (cotyledons) – monocot seeds have one and most other seed plants have two. Some cotyledons provide food for the plant embryo, as does endosperm in a monocot. Both types of seed contain epicotyls: shoots for upper stems and leaves; hypocotyls, which form lower stems; and radicles, which form roots.

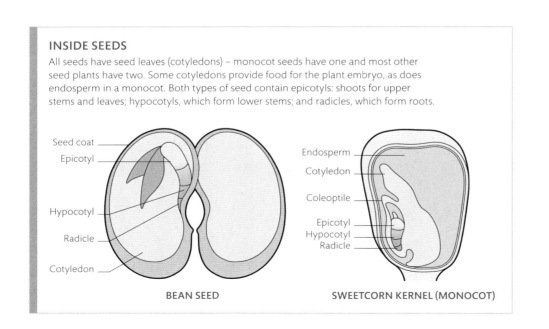

Seed coat
Epicotyl
Hypocotyl
Radicle
Cotyledon

BEAN SEED

Endosperm
Cotyledon
Coleoptile
Epicotyl
Hypocotyl
Radicle

SWEETCORN KERNEL (MONOCOT)

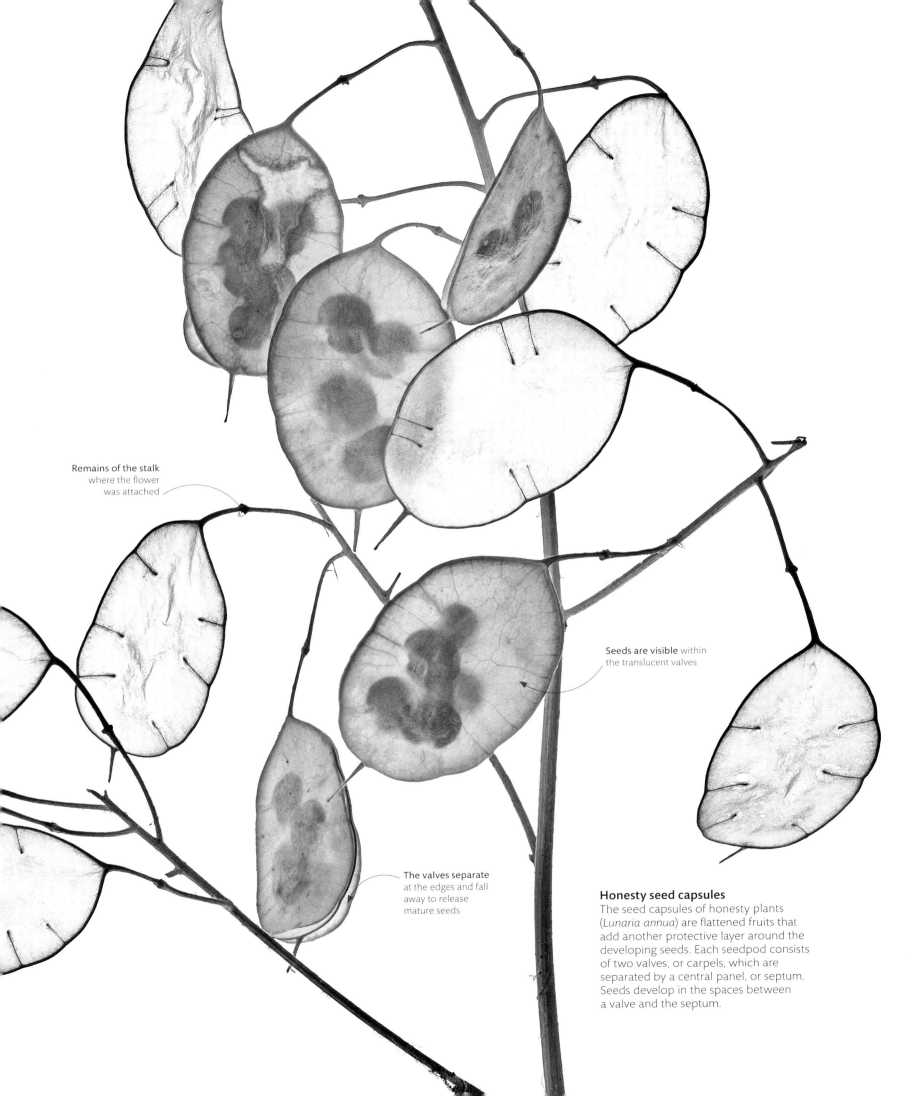

Remains of the stalk where the flower was attached

Seeds are visible within the translucent valves

The valves separate at the edges and fall away to release mature seeds

Honesty seed capsules
The seed capsules of honesty plants (*Lunaria annua*) are flattened fruits that add another protective layer around the developing seeds. Each seedpod consists of two valves, or carpels, which are separated by a central panel, or septum. Seeds develop in the spaces between a valve and the septum.

naked seeds

Developing without surrounding ovaries, the naked seeds of gymnosperms are exposed to the environment. Like those of flowering plants, naked seeds have coats, but they mature within cones instead of fruits. The most familiar are woody conifer cones, whose scales shield seeds as they ripen. Others, such as yews, produce single seeds that grow in fleshy cases.

Woody cones
Seed-bearing conifer cones come in a wide variety of shapes and sizes. Not all cones seem to match the size of the trees that produce them. Those of the giant sequoia are just 5–8 cm (2–3 in) long, although the tree reaches heights of 94 m (310 ft).

Three-pointed bracts, known as "rat tails", protrude from scales

Tightly closed scales peel away to release wide-winged seeds

PSEUDOTSUGA MENZIESII

CEDRUS ATLANTICA

Cones live for decades, and only release their seeds if affected by fire, squirrels, or beetles

The massive cones are 24–40 cm (9–15 in) long and weigh up to 5 kg (11 lb) when fresh

PINUS COULTERI

SEQUOIADENDRON GIGANTEUM

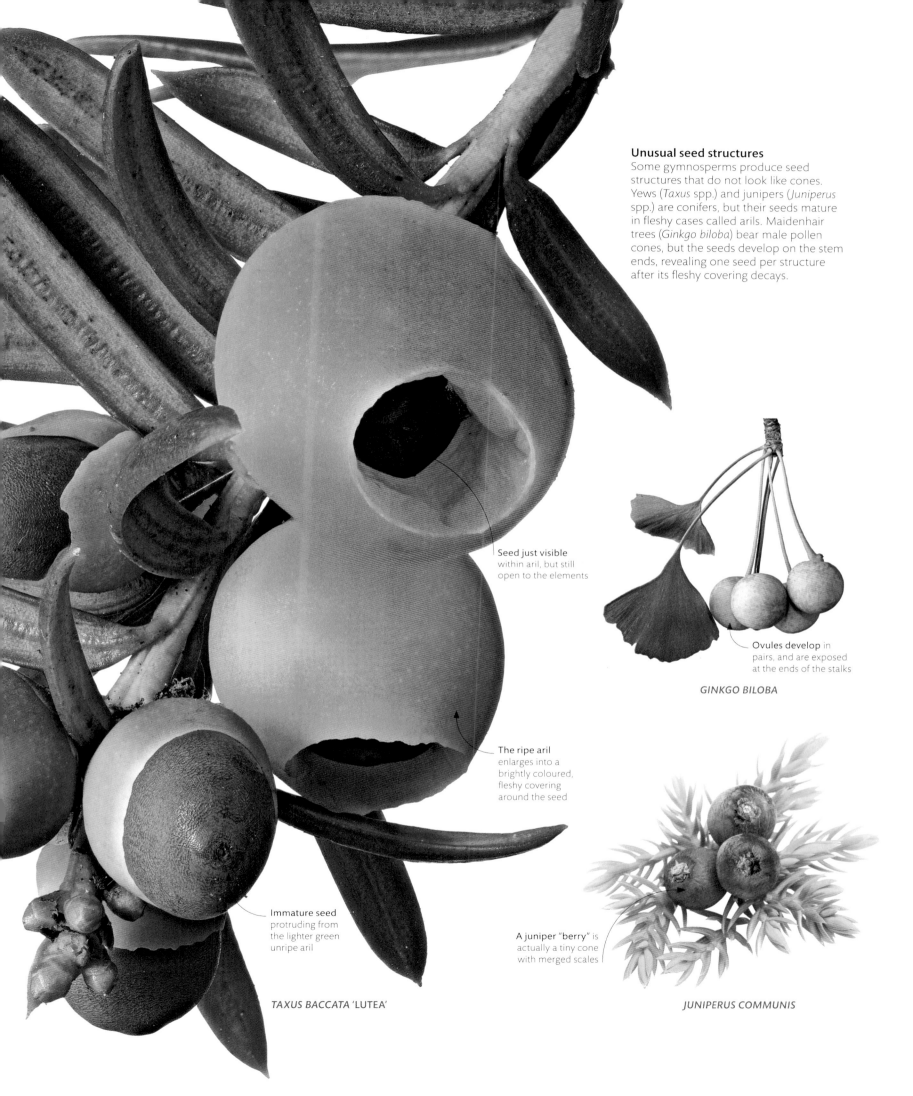

Unusual seed structures

Some gymnosperms produce seed structures that do not look like cones. Yews (*Taxus* spp.) and junipers (*Juniperus* spp.) are conifers, but their seeds mature in fleshy cases called arils. Maidenhair trees (*Ginkgo biloba*) bear male pollen cones, but the seeds develop on the stem ends, revealing one seed per structure after its fleshy covering decays.

Seed just visible within aril, but still open to the elements

Ovules develop in pairs, and are exposed at the ends of the stalks

GINKGO BILOBA

The ripe aril enlarges into a brightly coloured, fleshy covering around the seed

Immature seed protruding from the lighter green unripe aril

TAXUS BACCATA 'LUTEA'

A juniper "berry" is actually a tiny cone with merged scales

JUNIPERUS COMMUNIS

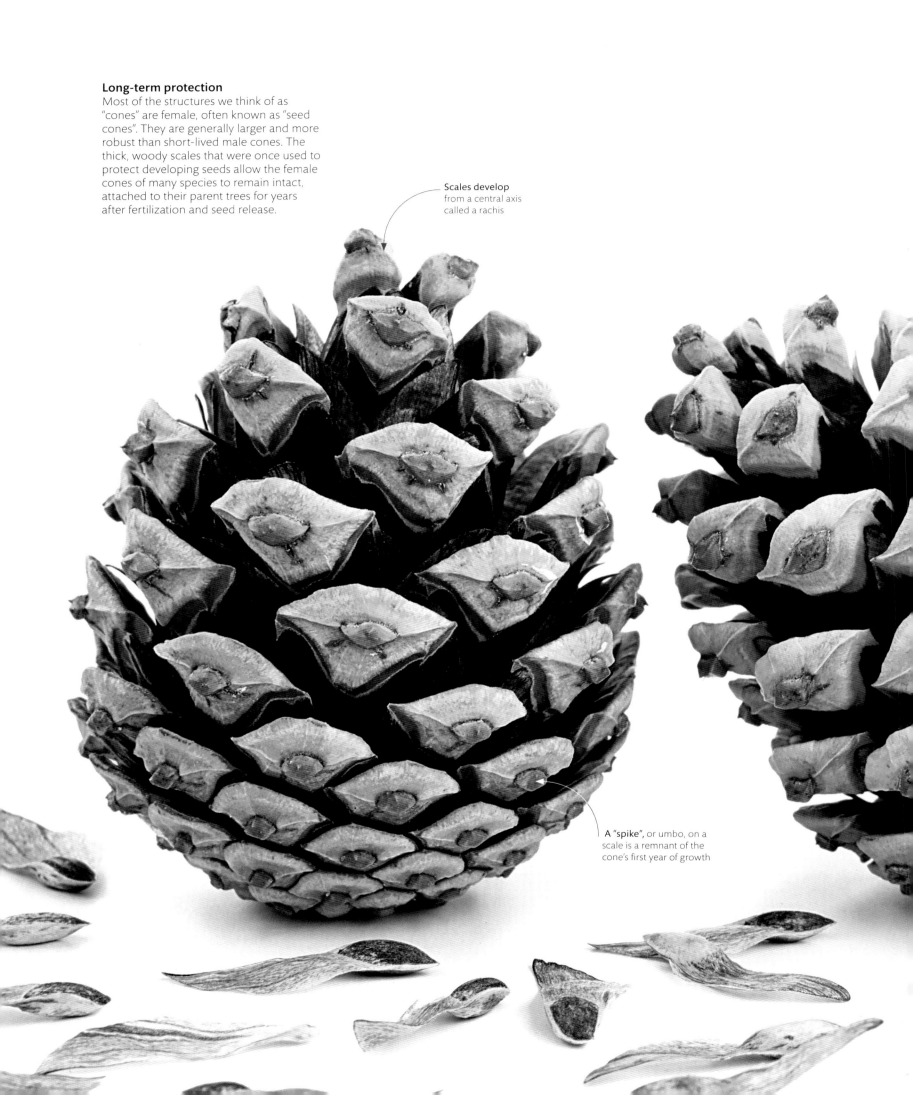

Long-term protection

Most of the structures we think of as "cones" are female, often known as "seed cones". They are generally larger and more robust than short-lived male cones. The thick, woody scales that were once used to protect developing seeds allow the female cones of many species to remain intact, attached to their parent trees for years after fertilization and seed release.

Scales develop from a central axis called a rachis

A "spike", or umbo, on a scale is a remnant of the cone's first year of growth

SEED DEVELOPMENT

Conifer seeds develop when pollen grains released by male cones fertilize ovules on the scales of female cones. The wind carries pollen to female cones, and it enters via a small opening called a micropyle. The pollen grains form tubes through which male gametes travel to fuse with female gametes in the ovules. Once fertilized, the ovule develops into an embryo surrounded by a seed coat and protected by scales.

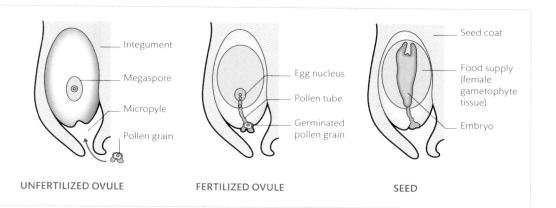

UNFERTILIZED OVULE
- Integument
- Megaspore
- Micropyle
- Pollen grain

FERTILIZED OVULE
- Egg nucleus
- Pollen tube
- Germinated pollen grain

SEED
- Seed coat
- Food supply (female gametophyte tissue)
- Embryo

inside seed cones

Gymnosperms have two distinct phases to their life cycles. Both male and female reproductive structures form in cones, where each generates sex cells, or gametes, that are haploid – meaning that the nucleus of each cell carries only one set of chromosomes. Each seed produced by the union of male and female gametes during fertilization is diploid: containing two sets of chromosomes. Many gymnosperm trees are diploid organisms, created by the fusion of two separate haploid cells.

An undetached seed remains lodged between scales

A fertilized ovule has developed into a seed, wedged between scales

Internal anatomy
Cutting through an unopened female cone shows how tightly pressed together the scales are, and how effectively the developing seeds are protected.

Lantern skeletons remain
on the plant months after
the seeds have dropped

enclosed seeds

Enclosed seeds develop within ovaries, which form fruit coverings that protect developing seeds as they ripen. This extra layer may also act as food to attract animals that aid in seed dispersal. Additional coverings vary from being quite hard, such as the layers surrounding coconut seeds, to being so fragile that, at first glance, they seem to have little purpose at all.

NUT OR SEED?

Botanically, a "nut" is defined as an enclosed seed with a hard outer shell that is indehiscent, meaning that it does not open naturally to release the seed. Although both sweet and horse chestnuts include the word in their names, neither qualifies as a nut. Both may be protected by formidable spiky coverings, but because the coverings eventually fall apart naturally, what is inside is classed as a seed with a seed coat. Other common misnomers include Brazil nuts and cashews, which both mature in dehiscent pods.

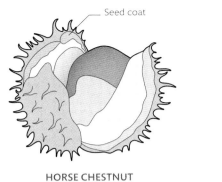

Seed coat

HORSE CHESTNUT

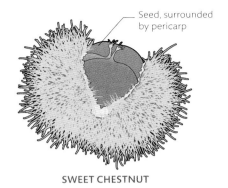

Seed, surrounded by pericarp

SWEET CHESTNUT

Each *Physalis* fruit contains multiple seeds

Chinese lanterns

Each of the attractive, paper-like coverings of the Chinese lantern (*Physalis* sp.) are a calyx – the sepals of a flower, which have fused and inflated to envelop a single berry-like fruit. What the "lanterns" lack in durability they make up for in other ways. While the fruits they surround may be edible, the calyx is toxic, and this characteristic, combined with its ability to provide protection against the weather, forms an effective enclosure.

Colourful tissues break down between the veins of the deteriorating calyx

The stem shrivels and bends as the plant deteriorates

Mature stems are brown and woody and carry fruits produced from the previous year

types of fruit

After a flower is pollinated, the ovules within the ovary develop into seeds. The ovary wall, or pericarp, forms a protective layer enclosing the seeds, which comprises the fruit. The way in which the pericarp develops determines the type of fruit, with some becoming fleshy and edible, while others are dry and largely inedible. In many fruits, the pericarp is differentiated into three layers: the skin, or epicarp; the flesh, or mesocarp; and the stone, or endocarp.

Flower to fruit
At the heart of each madrone (*Arbutus menziesii*) flower is an ovary. Once fertilized, this will develop into a simple fleshy fruit.

A green ovary sits in the centre of the flower, attached to the receptacle

Five petals fuse into an urn-shaped tube that largely encloses the flower parts

An urn-shaped corolla is common in flowers pollinated by bumblebees

Not all flowers develop to maturity; some abort at the bud stage

The colour changes from yellow to red when ripe, a signal that the fruits are at their sweetest

Arbutus fruits are popular with birds, who distribute the seeds far from the parent plant

Distinguishing features

Fleshy, strawberry-shaped fruits give the strawberry tree (*Arbutus unedo*) its name, but they are not actually strawberries. Each fruit is formed from the ovary within a single flower and is therefore a simple fruit. By contrast, a strawberry is an accessory fruit with a fleshy receptacle.

FRUITS AND FLOWER HEADS

Simple fruits come from a single flower with one ovary. Aggregate fruits are also the product of a single flower, but one with several ovaries. Multiple fruits derive from several close-set flowers, while accessory fruits incorporate tissues other than the pericarp.

Single ovary contains ovules

Stigma receives pollen

Petals attract pollinators

SIMPLE FRUIT (E.G. CHERRY)

Each ovary develops into fruitlet

Multiple ovaries

AGGREGATE FRUIT (E.G. RASPBERRY)

Flowers in inflorescence

Separate fruits fuse together

MULTIPLE FRUIT (E.G. PINEAPPLE)

Ovules inside ovary

Receptacle becomes part of fruit

ACCESSORY FRUIT (E.G. APPLE)

Pavement of fruit trees
This mosaic of a pomegranate tree, possibly dating back to the reign of Byzantine emperor Heraclius (c.575–641 CE), was part of a pavement decoration from the Great Palace of Constantinople.

plants in art

ancient gardens

The first gardens were created by the earliest societies of the Middle East, when the need for self-sufficiency led people to enclose plots of land next to their homes. Over time, the practical function of the garden was superseded by people's desire to enhance their surroundings, with the emerging ruling classes using gardens to enjoy their leisure time and reinforce their status.

Glimpses of ancient gardens and plants are found in archaeology, literature, and art across the ancient world.

The first large-scale formal gardens were built by the Emperors of ancient Mesopotamia, home to the legendary Hanging Gardens of Babylon. These gardens often combined elaborate irrigation systems and stone landscaping with formal planting of trees and exotic plants aquired on foreign campaigns.

The ancient Egyptians created gardens for both secular and religious purposes; temples often had gardens in their compounds, where symbolic herbs and vegetables, and plants used in rituals were

grown. The Egyptians also cultivated many types of flowers that they used in festive garlands and for medicinal purposes.

Domestic pleasure gardens were rare in ancient Greece. Relatively simple gardens were closely associated with religion, and the trees and plants grown in them were associated with particular deities.

Heavily influenced by Egypt and Persia, garden design and horticultural techniques became highly advanced in ancient Rome. From the town houses of Pompeii to Rome's imperial palaces, gardens were places for relaxation and escape, and often featured art and objects with religious and symbolic meaning.

Timeless garden
A fresco from the Villa of Livia, near Rome, built for the wife of the Roman Emperor Augustus, depicts a garden that is both naturalistic and borne of fantasy. By fruiting and flowering at the same time, the trees and bushes serve to convey the fecund "perpetual spring" of the glorious reign of the Emperor.

If you have a garden and a library,
you have everything you need.

MARCUS TULLIUS CICERO, *TO VARRO, IN AD FAMILIARES IX, 4*

Ripe blackberries
form at the end of
the panicle

**RIPENING
BLACKBERRY**

Thorns help protect
the berries from being
eaten by predators

Stamens begin to
wither as fruitlets
emerge after
fertilization

Blackberry fruits
Blackberry (*Rubus* sp.) bushes produce
long panicles (see p.216), whose branches
end in flower buds. The flowers at the end
of the stem usually bloom and ripen
before the other flowers, so that the fruits
develop at different times on just one
section of a blackberry bush.

Each "berry" is formed of several tiny fruitlets

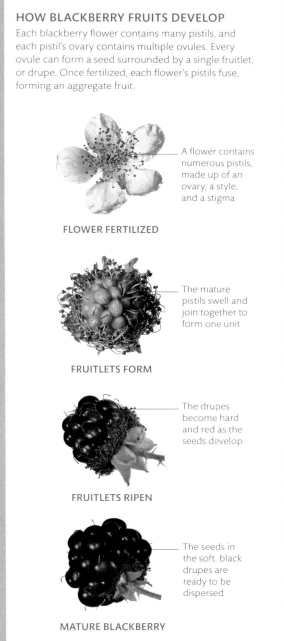

HOW BLACKBERRY FRUITS DEVELOP

Each blackberry flower contains many pistils, and each pistil's ovary contains multiple ovules. Every ovule can form a seed surrounded by a single fruitlet, or drupe. Once fertilized, each flower's pistils fuse, forming an aggregate fruit.

A flower contains numerous pistils, made up of an ovary, a style, and a stigma

FLOWER FERTILIZED

The mature pistils swell and join together to form one unit

FRUITLETS FORM

The drupes become hard and red as the seeds develop

FRUITLETS RIPEN

The seeds in the soft, black drupes are ready to be dispersed

MATURE BLACKBERRY

flower to fruit

When flowers appear in late spring or early summer, they mark the first stage of fruit formation. The next stage occurs when a grain of pollen from a plant of the same species lands on a flower's stigma. This produces a pollen tube that travels through its style. This "tunnel" gives the pollen's nucleus access to the flower's ovule-filled ovary (see pp.184–185), where it fuses with an ovule nucleus, fertilizing it. Fertilization signals the end for the flower, but as its petals wither and fall, all fertilized ovules transform into seeds, and their surrounding ovaries swell and ripen into fruit.

succulent fruit

Fruit from single flower with single ovary

BERRY
Solanum betaceum

Berry with segmented flesh

HESPERIDIUM
Citrus x limon

Unsegmented berry with tough rind

PEPO
Cucumis metuliferus

Pseudofruit, as flesh does not form from ovary and seeds are achenes

HIP
Rosa rugosa

dry fruit

Single-seeded fruit derived from whole ovary

Single-seeded fruits, each derived from one carpel of an ovary

CYPSELA
Taraxacum sp.

ACHENE
Fragaria x ananassa

Winged achene with wing that surrounds fruit

SINGLE SAMARA
Ulmus glabra

Paired achenes derived from flower with two carpels

DOUBLE SAMARA
Acer tataricum subsp. ginnala

fruit anatomy

Essentially, fruit classification is based on a handful of characteristics and one of the most important is texture. Succulent fruits are eaten and dispersed by animals, whereas dry fruits rely on wind, gravity, or animal fur for dispersal. Although inspecting a fruit, and ideally the flower it is derived from, should make everything clear, botanical classifications can still be surprising. A cucumber, for example, is classified as a berry, while a strawberry is not.

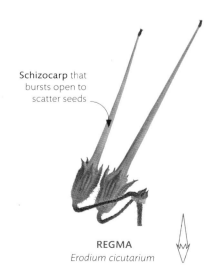

Schizocarp that bursts open to scatter seeds

REGMA
Erodium cicutarium

Seeds enclosed within leathery core

POME
Malus x domestica

Seeds are enclosed within woody pit

DRUPE
Prunus persica

From a single flower with multiple carpels

AGGREGATE
Rubus idaeus

Small fruits fuse together into one unit

MULTIPLE
Maclura pomifera

Hard-walled fruit that does not split open to release the seed

NUT
Corylus colurna

Similar to achenes, but outer coat is fused to the seed

CARYOPSIS
Zea mays

Derived from one carpel and split along two seams

LEGUME
Lathyrus odoratus

May cluster together and each splits along one seam

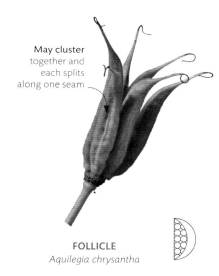

FOLLICLE
Aquilegia chrysantha

Multiple chambers distinguish capsules from other dry fruits

CAPSULE
Rhodophiala bifida

Capsule develops open pores to shed seeds

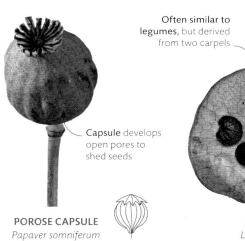

POROSE CAPSULE
Papaver somniferum

Often similar to legumes, but derived from two carpels

SILIQUE
Lunaria sp.

Fruit splits into one-seeded pieces (mericarps)

SCHIZOCARP
Trachyspermum ammi

Musa sp.

the banana

The long, slender fruits of a banana plant are actually berries. Store-bought bananas are seedless, but in the wild these fruits are full of seeds hard enough to crack a tooth. There are 68 different banana species, all belonging to the genus *Musa* and hailing from tropical Indomalaya and Australia.

The species in the *Musa* genus differ greatly in height, from the diminutive *Musa velutina*, which rarely grows to be much more than 2 m (6½ ft) tall, to the monstrous *Musa ingens*, which regularly reaches 20 m (66 ft). Despite their appearance, bananas are not trees: they do not produce any wood and what looks like a tree trunk is actually the hardened, tightly packed bases of their long, tropical leaves. Indeed, bananas are the largest herbaceous plants on the planet.

There are two major flowering forms, upright and drooping. In the upright form, the flowers point towards the sky and are pollinated mainly by birds; in the drooping form, the flowers point towards the ground and are pollinated largely by bats. In both forms, the flowers are produced on a spike-like inflorescence, and each whorl of flowers is protected by large, often showy bracts.

In the wild, bananas only form fruits if they are pollinated. The fruits start out greenish in colour but change hue as they ripen. Not all bananas end up yellow: some species produce bright pink fruits. If animals harvest the fruits too early, the banana seeds may not be viable. The colour change occurs only when the seeds inside are ready, and it signals to animals that the fruits are fit to eat. Researchers have found that, upon ripening, some bananas fluoresce under ultraviolet (UV) light. This may help animals that can detect UV to find ripe fruits more easily.

Flowers and fruit
Banana flowers are usually yellow or cream in colour, and tubular in shape. Birds and bats are the main pollinators of wild bananas. Cultivated bananas produce fruits without pollination, so they are all seedless.

A banana stem (there is only one stem per banana plant) can produce as many as 200 fruits

Banana bunch
Bananas were first cultivated over 7,000 years ago. Today, cultivated bananas all over the world are descended from only two species, *Musa acuminata* and *Musa balbisiana*. Banana farming uses mostly clones of only a few banana varieties, making these plants far more susceptible to disease than wild ones.

Hairs on a flower head signal developing or fertilized spikelets

The dried-out, yellowed inflorescence retains spikelets after the seeds are released

Equipped for dispersal
As blue grama grass seeds ripen, the flower head curves, opening the spikelets and giving the plant its nickname "eyelash grass". Each seed has three bristly awns that are capable of hooking onto fur, clothing, or feathers.

seed distribution

Seeds and spores are the travellers of the botanical world. No matter what type or size they are, seeds have one crucial mission: to transport the genetic material needed to create a new plant. For some species, this is as simple as falling onto fertile ground from a parent plant. For others, it means hitching a ride on whatever is available – wind, water, birds, insects, animals, humans – or even being ingested and deposited many miles away from the plant on which they formed.

The terminal (top) inflorescence is in the best position to release seeds on the wind

Each panicle supports 1–3 main flowering branches, usually ending with a terminal inflorescence

Multiple dispersal methods

Launched from stalks up to 30 cm (12 in) high, seeds of blue grama grass (*Bouteloua gracilis*) have many opportunities to disperse. The seeds catch passing breezes to float several metres away, and some will germinate where they land. They are also eaten by herbivores and grow where they are excreted. Seeds that cling to animal fur or bird feathers travel even further from their parent plant.

Seeds are expelled from the ends of spikelets when the seedhead matures

Each blue grama inflorescence contains up to 130 spikelets

Takeaway advertising
The brightly coloured berries of Tasman flax-lilies (*Dianella tasmanica*) are easy to spot from the ground or the air, making them very attractive to birds. By dangling its high-carbohydrate fruits at the ends of long stems, the plant also makes it easy for birds to reach them as they fly past.

Each berry contains five black seeds that pass through birds' digestive systems intact

Vivid colours, such as reds and purples, are known to attract wild birds

Germination boost
Seeds of rowan trees (*Sorbus aucuparia*) dispersed in bird droppings germinate faster than non-digested seeds. This might be due to some chemicals being removed during the digestive process.

A single berry at the end of a stem is easy to pluck

dispersed by diet

Some plant species are much more successful than others at distributing their seeds over wide areas. One of the most effective ways for the plants to accomplish this is by feeding birds. Although many other animals also spread seeds through their droppings, airborne birds can cover a much wider territory, increasing the likelihood of dropping seeds from a recent meal far away from their source.

FORGOTTEN FOOD

Squirrels are among the most famous "wild gardeners" due to their habit of burying, then losing track of, winter nut stashes. By caching food underground for cold-weather meals, these mammals inadvertently plant numerous trees, especially hazels and oaks, whose saplings reveal the lost storage locations the following spring. Buried nuts and acorns have much higher germination rates than those that just fall to the ground from the branch.

Acorn shells protect seed underground

Cupule (cup) formed from bracts at flower base

STRANDZHA OAK (*QUERCUS HARTWISSIANA*)

Green, unripe berries are inconspicuous

TASMAN FLAX-LILY
(*DIANELLA TASMANICA*)

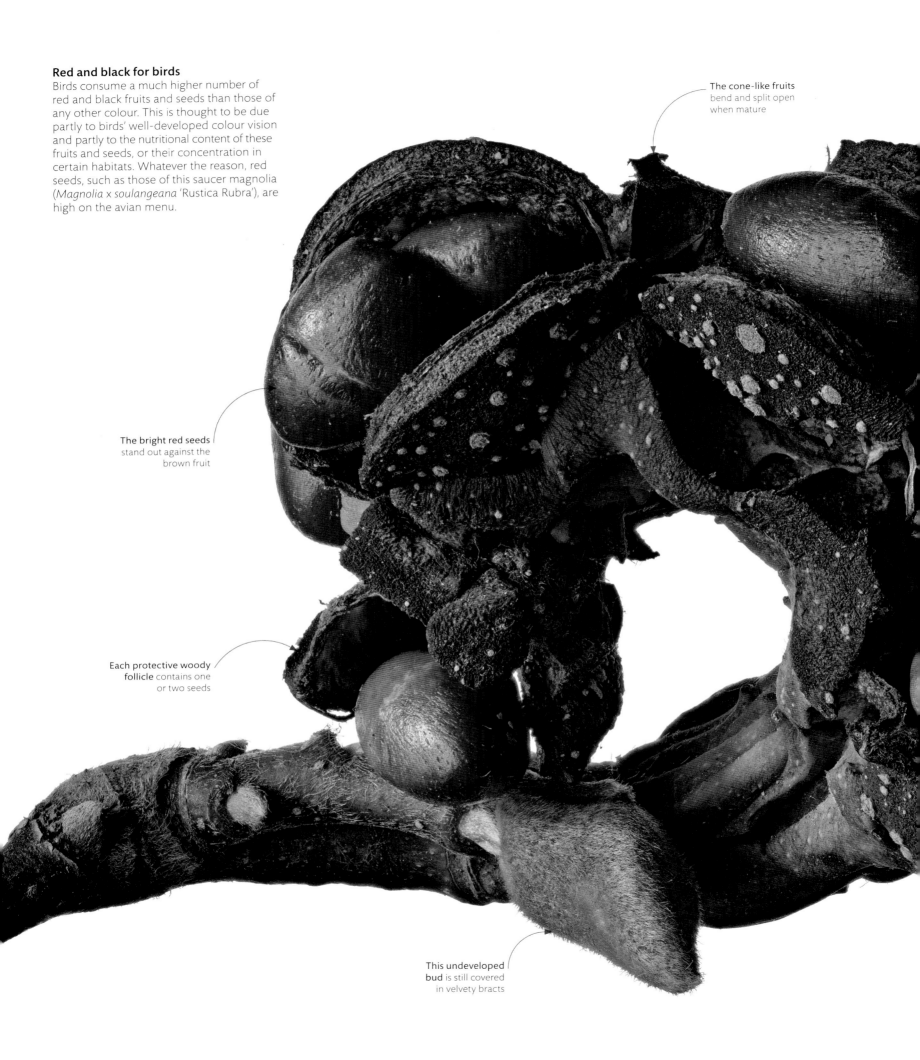

Red and black for birds

Birds consume a much higher number of red and black fruits and seeds than those of any other colour. This is thought to be due partly to birds' well-developed colour vision and partly to the nutritional content of these fruits and seeds, or their concentration in certain habitats. Whatever the reason, red seeds, such as those of this saucer magnolia (*Magnolia* x *soulangeana* 'Rustica Rubra'), are high on the avian menu.

The cone-like fruits bend and split open when mature

The bright red seeds stand out against the brown fruit

Each protective woody follicle contains one or two seeds

This undeveloped bud is still covered in velvety bracts

BIRD OF PARADISE
(*STRELITZIA REGINAE*)

TRAVELLER'S PALM
(*RAVENALA MADAGASCARIENSIS*)

The orange arils appeal to monkeys, while the black seeds attract birds

Colour and seed distribution
Both mammals and birds consume seeds and distribute them in droppings or by hiding them away, but their colour preferences vary. Research worldwide shows that birds prefer red and black seeds, while mammals eat mainly orange, yellow, and brown ones. When certain creatures only see or prefer certain shades, plants may have adapted, sometimes simply by adding vibrant arils (hairy coverings) to otherwise drab seeds.

Striking blue arils decorate the blue seeds of the traveller's palm, attracting lemurs, which can only see blue and green

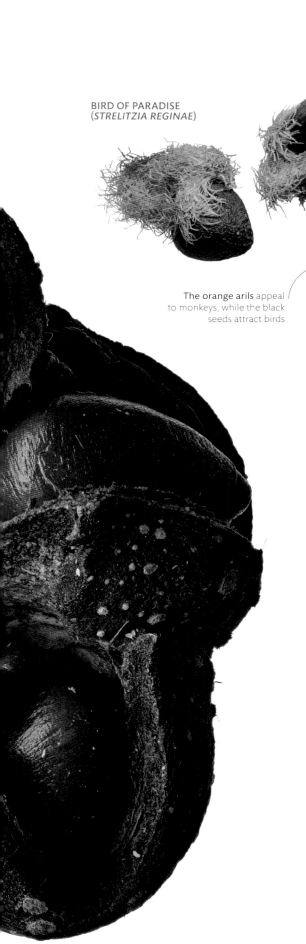

colourful seeds

Like the fruits that bear them, seeds come in a myriad of colours, with protective seed coats ranging from plain black to dazzling reds, oranges, and blues. Although we know that light-coloured seeds often contain more moisture than dark ones, the reasons why some seeds have such vivid pigments are still unclear. What is certain, however, is that particular colours seem to be preferred by certain animals.

TOXIC RICIN SEEDS
The main function of any seed coat is protection, sometimes protection from being consumed. Toxic seeds contain some of the most deadly substances on Earth. Ricin, which is found in castor beans, is so poisonous that just four beans are enough to kill a human. Because the colour of castor bean varies from solid white to mottled red and black, many animals and birds die from ingesting it.

CASTOR OIL PLANT
(*RICINUS COMMUNIS*)

Mottled colouring, similar to that of edible lima beans

The caruncle is a spongy outgrowth full of sugars that attracts ants

a. *Malus oxymela acida*, Saurer Holtzapfel. b. *Ma=*
lus sylvestris fructu rubro minore, Pomme sauvage,
Holtzapfel. c. *Malus sylvestris fructu rotundo viridi*,
grüne Holtzapfel. d. *Malus Persica flore pleno*. e. *Malus Persica*
Sti Laurentii dicta. f. *Malus Persica minor*, Pesche petit, Pfirsig. g. *Ma=lus*
Persica major molle carne Pfirsigapfel. h. *Malus Persica magna*, Bockater Pfirsig.

art and science

The 18th century is often called the Golden Age of botanical painting, and the illustrations of the botanic artist Georg Dionysius Ehret represent a great meeting of art and science. Ehret matched Carl Linnaeus's groundbreaking approach to the naming and classification of plants and animals with a clear, precise, and beautiful style of plant illustration.

One of the most influential botanical illustrators of all time, German-born Ehret (1708–70) was the son of a gardener who taught him about nature. Ehret's talent for drawing, eye for detail, and growing knowledge of plants led him to produce botanical artworks that brought him to the attention of some of the world's leading scientists and influential patrons.

Ehret first collaborated with the famous Swedish botanist and taxonomist Carl Linnaeus on *Hortus Cliffortianus* (1738), a catalogue of rare plants found on the estate of George Clifford, the governor of the East India Company. Under Linnaeus's direction, Ehret documented every part of the plants in beautiful, yet scientifically accurate and detailed artworks, in a style that become known as the Linnaean style of botanical illustration.

Ehret went on to illustrate most of the important botanical publications of the day, and produced numerous illustrations for collectors and institutions, including the Royal Botanic Gardens at Kew.

Pineapple (*Ananas sativus*)
Ehret's work included pencil, ink, and watercolour studies of plants from around the world, such as this pineapple, which had recently arrived in London to be studied at the Chelsea Physic Garden, one of the oldest botanic gardens in Britain.

Details of *Malus* fruit and flowers
These illustrations of apple and peach flowers and fruit are mezzotint engravings that were coloured by hand. Ehret was commissioned by the Dutch apothecary Johann Wilhelm Weinmann to illustrate *Phytanthoza iconographia*, but he completed only half the illustrations due to the low fee paid by Weinmann.

66 The genius of Georg Dionysius Ehret was the dominant influence in botanical art during the middle years of the eighteenth century. 99

WILFRID BLUNT, *THE ART OF BOTANICAL ILLUSTRATION*, 1950

UNICORN FRUIT

Some of the largest hitch-hiking seedheads are produced by plants known as devil's claws, but the spiky pods don't emerge straight away. The woody seed capsule of *Ibicella lutea* forms within a large, horn-like fruit, earning it the name of "unicorn plant".

***IBICELLA LUTEA* SEED CAPSULE**

The elongated primary claws curve upwards to increase the chances of latching onto animals

Shorter secondary claws sometimes form between the primary claws

Barbs cover all the claws of the seedhead, and hook on to the feet and legs of passing animals

animal couriers

To avoid overcrowding, plants have to disperse seeds over as wide an area as possible. Some plants use wind and water and others expel their seeds via "exploding" seedheads. Many plants use animals that share their environments to scatter their seeds. "Sticky" seed capsules covered in hooks, barbs, and painful spines latch on to hide and hoof, and only release their seeds when the capsule is rubbed off, crushed, or torn open, often many kilometres away.

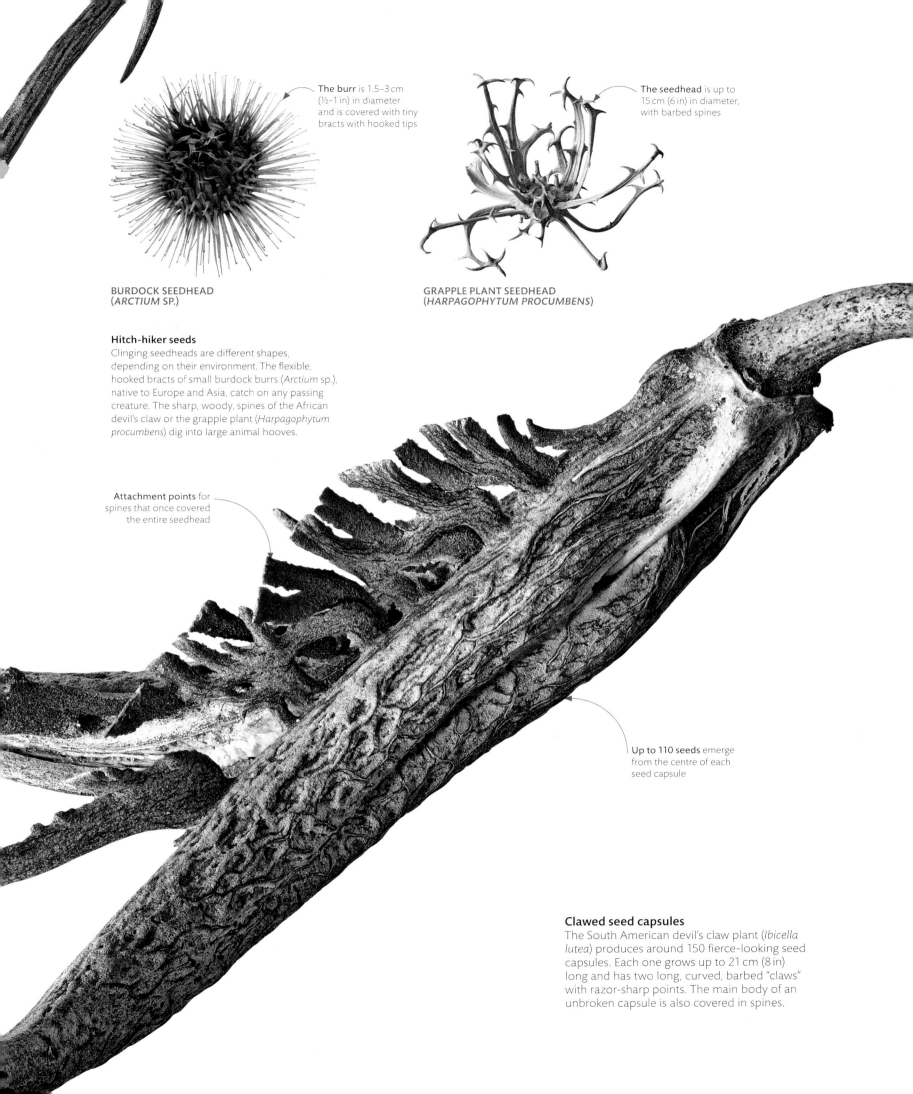

The burr is 1.5–3 cm (½–1 in) in diameter and is covered with tiny bracts with hooked tips

The seedhead is up to 15 cm (6 in) in diameter, with barbed spines

BURDOCK SEEDHEAD
(*ARCTIUM* SP.)

GRAPPLE PLANT SEEDHEAD
(*HARPAGOPHYTUM PROCUMBENS*)

Hitch-hiker seeds

Clinging seedheads are different shapes, depending on their environment. The flexible, hooked bracts of small burdock burrs (*Arctium* sp.), native to Europe and Asia, catch on any passing creature. The sharp, woody, spines of the African devil's claw or the grapple plant (*Harpagophytum procumbens*) dig into large animal hooves.

Attachment points for spines that once covered the entire seedhead

Up to 110 seeds emerge from the centre of each seed capsule

Clawed seed capsules

The South American devil's claw plant (*Ibicella lutea*) produces around 150 fierce-looking seed capsules. Each one grows up to 21 cm (8 in) long and has two long, curved, barbed "claws" with razor-sharp points. The main body of an unbroken capsule is also covered in spines.

Ideal launchpad
The tree of heaven (*Ailanthus altissima*) is so named because it quickly achieves heights of 24 m (80 ft) or more – perfect for wind-borne seed dispersal. A single tree can launch a million seeds a year – and because of this, it can become an invasive species in non-native lands.

Single-seed samaras break off both individually and in whole clusters

The rigid, seam-like edge stabilizes the samara as it spins

The semi-transparent samara wing allows seeds to glide up to 90 m (300 ft) away

The division line of the double samara, where the seeds will separate from each other

The pericarp wall extends and elongates to form a thin, membrane-like wing

The pedicel attaches the seed to the tree and separates from the stem when it matures

The pericarp layer encloses a single seed

seeds with wings

Putting enough distance between the parent plant and its seeds is vital for trees, which can easily grow too close together. The seeds of maples, sycamores, and many other species have evolved wing-like extensions that enable them to catch the breeze and travel much further. Whether they glide, spin, or float, all winged seeds, called samaras, take their chances on the air.

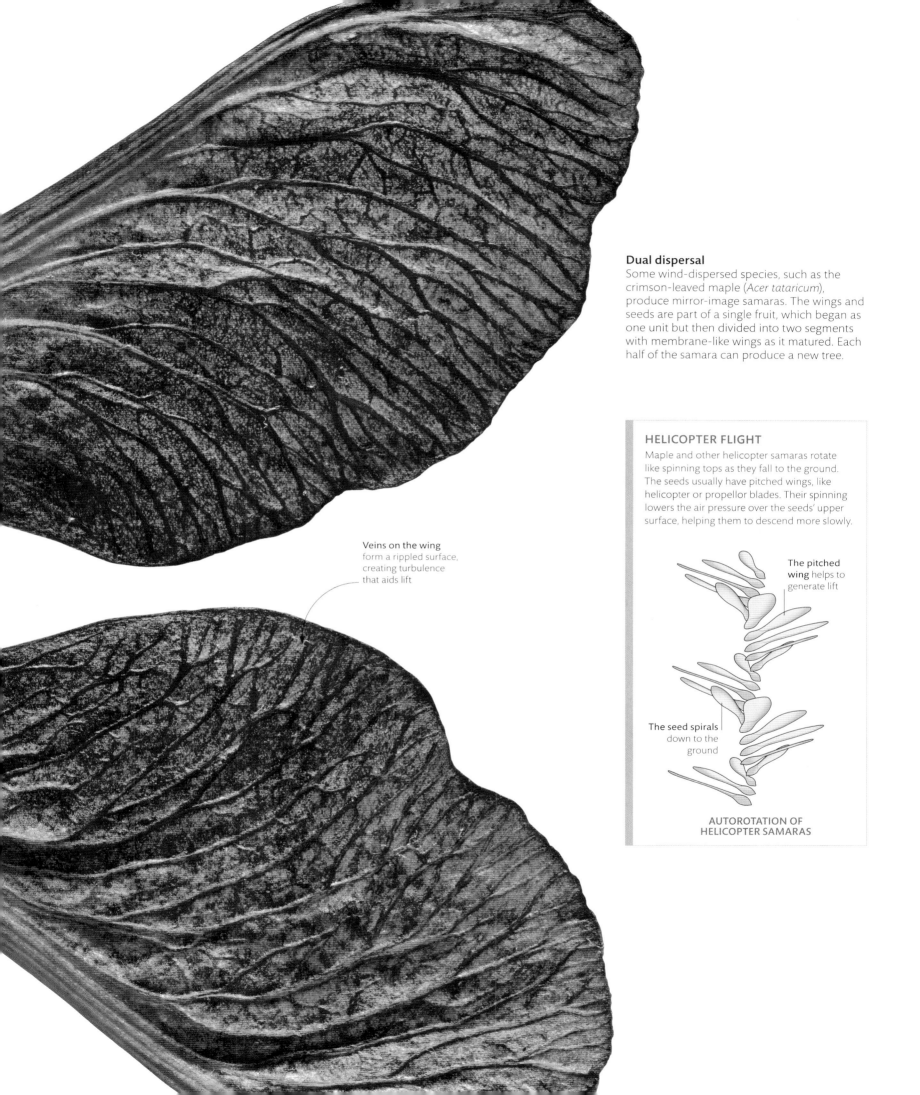

Dual dispersal
Some wind-dispersed species, such as the crimson-leaved maple (*Acer tataricum*), produce mirror-image samaras. The wings and seeds are part of a single fruit, which began as one unit but then divided into two segments with membrane-like wings as it matured. Each half of the samara can produce a new tree.

HELICOPTER FLIGHT

Maple and other helicopter samaras rotate like spinning tops as they fall to the ground. The seeds usually have pitched wings, like helicopter or propellor blades. Their spinning lowers the air pressure over the seeds' upper surface, helping them to descend more slowly.

Veins on the wing form a rippled surface, creating turbulence that aids lift

The pitched wing helps to generate lift

The seed spirals down to the ground

AUTOROTATION OF HELICOPTER SAMARAS

Taraxacum sp.

the dandelion

Named after their jagged leaf edges (from the French *dent-de-lion*, meaning "lion's tooth"), dandelions are loved by children for their puffy seedheads and despised by gardeners for spoiling lawns. Most species are native to Eurasia but have been widely spread by human help, so these botanic travellers are now found in almost all the world's temperate and subtropical regions.

Dandelion is a catch-all term for about 60 species belonging to the genus *Taraxacum*; the one most often encountered is the common dandelion (*Taraxacum officinale*). Key to the success of this adaptable plant is its reproductive strategy. Dandelions are among the first plants to bloom in spring, making them an important resource for insects when little else is in flower, and helping to ensure they are pollinated. Having flowered in spring, they will often bloom again in autumn. Crucially, though, the dandelion can produce seeds without being pollinated. When this occurs, the seeds grow into clones of the parent plant (apomixis). With each flower head having up to 170 seeds, and as a single plant is able to produce a total of more than 2,000 seeds, there is a high chance that more than one will grow into an adult plant.

With light, feathery parachutes, dandelion seeds are great aerial travellers. Most land fairly close to the parent plant, but some are blown by the wind or caught on updraughts of warm air and are then transported long distances.

Feathery parachutes
Each dandelion seed is attached by a stalk to a disc of radiating, feathery filaments that form a parachute-like structure.

Early bloomer
A dandelion's flower head is actually a composite of many individual flowers.

The spherical seedhead falls apart once it is ripe, and each single-seeded fruit drops away

The cluster of flowers provides the perfect landing pad for bumblebees and butterflies

Elaborate outer petals help to attract insect pollinators to the nectar-rich flowers

Strength in numbers
Scabious flowers are small and are gathered together in pincushion-like flower heads. After the flowers have been pollinated by insects, the petals drop, revealing a spherical head of papery fruits.

seeds with parachutes

Many plants use the wind to distribute their seed. In open environments, such as meadows and prairies where trees are few and far between, many plants do this as seeds can travel for long distances. To provide lift, wind-blown seeds need a sail or, in the case of this scabious, a parachute. Plants with wind-dispersed seeds often hold their flowers up high, so that when the seeds form, they can catch a breeze.

Spiky awns (bristles) catch in plants or on the ground, ending the seed's journey away from its parent plant

THE COMPLETE PACKAGE
Each seed contains an embryonic plant complete with a juvenile root (radicle) and shoot (hypocotyl), nestled within one or more cotyledons. These contain nutrients that fuel the plant as it germinates.

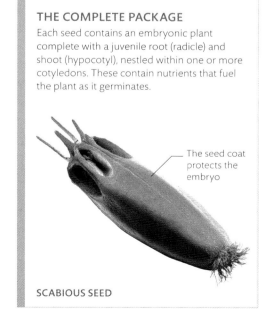

The seed coat protects the embryo

SCABIOUS SEED

Paper moons
The inflorescences of scabious (*Scabiosa stellata*) contain many small flowers, each of which produces a single seed. Every seed has five, spine-like bristles called awns, and is surrounded by papery bracts that form a parachute. The wind blows the parachute and the awns catch on the ground.

Papery bracts surround each fruit, giving them lift and helping them to hitch a ride on a breeze

Silky hairs catch the wind, transporting achenes far from the parent plant

Heading separate ways
Some flowers develop a single fruit, but those of *Clematis integrifolia* produce numerous one-seeded fruits called achenes. This is an advantage because the many fruits spread in various directions, expanding the plant's territory. Each achene has a tail covered in hairs that readily drifts on the wind.

silky seeds

To be transported by the wind, seeds need special adaptations. Some use silky hairs for the purpose, and in cotton (*Gossypium*) and poplar (*Populus*), the fruits are filled with a mass of hairs that are released with the seeds, aiding their dispersal. In other plants, the seed hairs develop into elaborate wings or parachutes. Hairs also help to fasten the seed to the surface on which it eventually lands, where it may be possible to germinate.

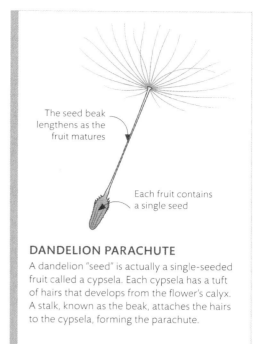

The seed beak lengthens as the fruit matures

Each fruit contains a single seed

DANDELION PARACHUTE
A dandelion "seed" is actually a single-seeded fruit called a cypsela. Each cypsela has a tuft of hairs that develops from the flower's calyx. A stalk, known as the beak, attaches the hairs to the cypsela, forming the parachute.

A hairy achene tail develops from the flower's old style and stigma

Numerous achenes remain attached to a central receptacle until they ripen and fall away

Cluster of carpels
At the heart of a clematis flower there are numerous carpels, each of which bears a single seed. These develop into achenes, and at maturity, the cluster of achenes breaks apart.

The long stem holds the seedhead aloft, so the wind is more likely to catch the fruits

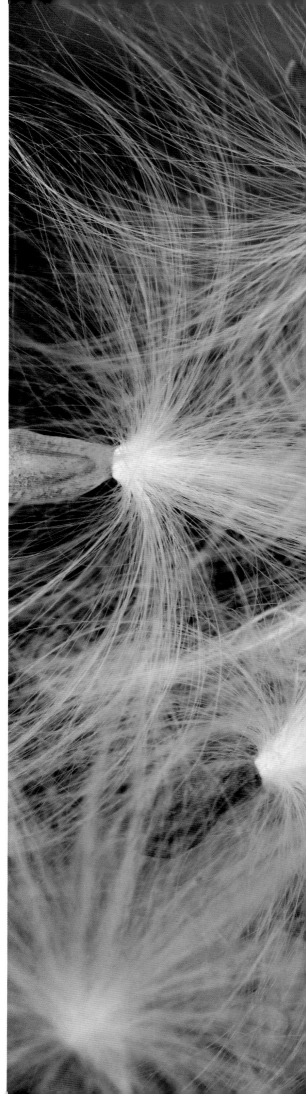

Asclepias syriaca

common milkweed

Found throughout much of eastern North America, milkweed is probably best known for being the larval food source of the monarch butterfly. At one time, milkweed was grown on a large commercial scale, and its "silk" was harvested as stuffing for pillows, mattresses, and even life jackets.

Milkweed flowers form umbrella-shaped inflorescences. Sweetly scented and ranging from light pink to near purple, individual flowers comprise five reflexed petals and five nectar-filled hoods. Unlike flowers that dust pollinators with copious amounts of pollen, milkweed blooms package their pollen into sticky sacs known as pollinia. The pollinia are held in grooves, called stigmatic slits, on either side of each hood. An insect arriving to drink nectar struggles to grip the flower's smooth surface and may accidentally slip a leg or two into the slits. The pollinia stick to the insect's legs and transfer to another flower when the insect gorges on the nectar of a neighbouring milkweed. This pollination strategy favours larger insects; some bees and smaller insects, however, get stuck in the slits or lose limbs in their efforts to escape. The fruits that form after pollination start off as tiny

green buds, then swell into large, seed-packed capsules known as follicles. Each flat brown seed in a follicle is adorned by a tuft of silky filaments called a coma.

The sap of milkweed contains toxins to deter mammal herbivores. Despite the toxicity, milkweed provides ample food in the form of nectar and leaves for a range of insects, including the monarch butterfly, that have adapted to its toxic defences. The fall in monarch numbers is partly linked to the milkweed's decline due to habitat loss and increasing use of herbicides. Changing the fortunes of the plant could revive butterfly numbers.

Silky seeds
New uses for milkweed silk – as thermal insulation for outdoor clothing, acoustic padding in vehicles, and an absorber of oil spills – may herald a renaissance for commercial milkweed farming.

Silky filaments are lightweight, hollow, and have a waxy, waterproof coating

Milkweed seeds are dispersed by wind, borne on their buoyant silk "parachutes"

Milkweed follicle
The seed capsules, or follicles, are 8–10 cm (3–4 in) long and covered with soft prickles and short woolly hairs. Once mature, the follicles split along the side to release the seeds.

HOW SEEDS GERMINATE

Most seeds contain an embryo and a food source to kickstart growth. At germination, roots emerge first, to anchor the seedling, followed by leaves. In most flowering plants, such as beans, a pair of seed leaves (cotyledons) are first to appear; these contain food transferred from the seed. Monocots have only one cotyledon, which may remain within the seed.

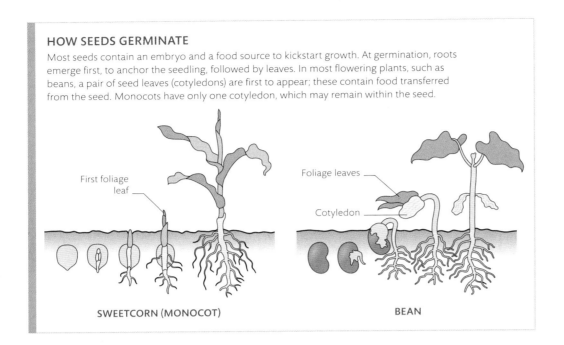

First foliage leaf

Foliage leaves

Cotyledon

SWEETCORN (MONOCOT)

BEAN

Shaken apart

The papery capsule fruits of the regal lily (*Lilium regale*) contain numerous winged seeds. When the fruits are dry, they split open and shed their seeds – with the help of the strong winter winds that sweep through the precipitous valleys of the lily's natural habitat in western China.

Lily capsules have three chambers, each containing circular, coin-like seeds

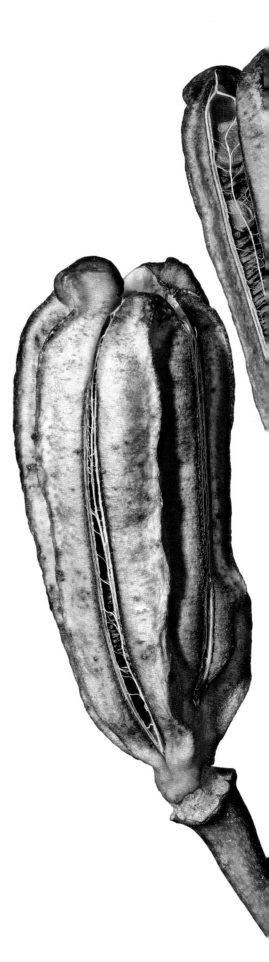

pods and capsules

Fleshy fruits are perhaps the best known as they are the ones we eat, but dry fruits are not uncommon. They include pods, capsules, achenes, follicles, and schizocarps, and they may split open to release their seeds (dehiscent) or remain intact to be distributed with the seed inside (indehiscent). Without a juicy coat to attract animals, dry fruits rely on other methods to disperse their seeds. Wind dispersal is widespread, but seeds can also cling to animal fur or fall to earth.

Valves or openings in lily fruit reveal two columns of seeds

The capsules dry out, shrink, and split open in late summer

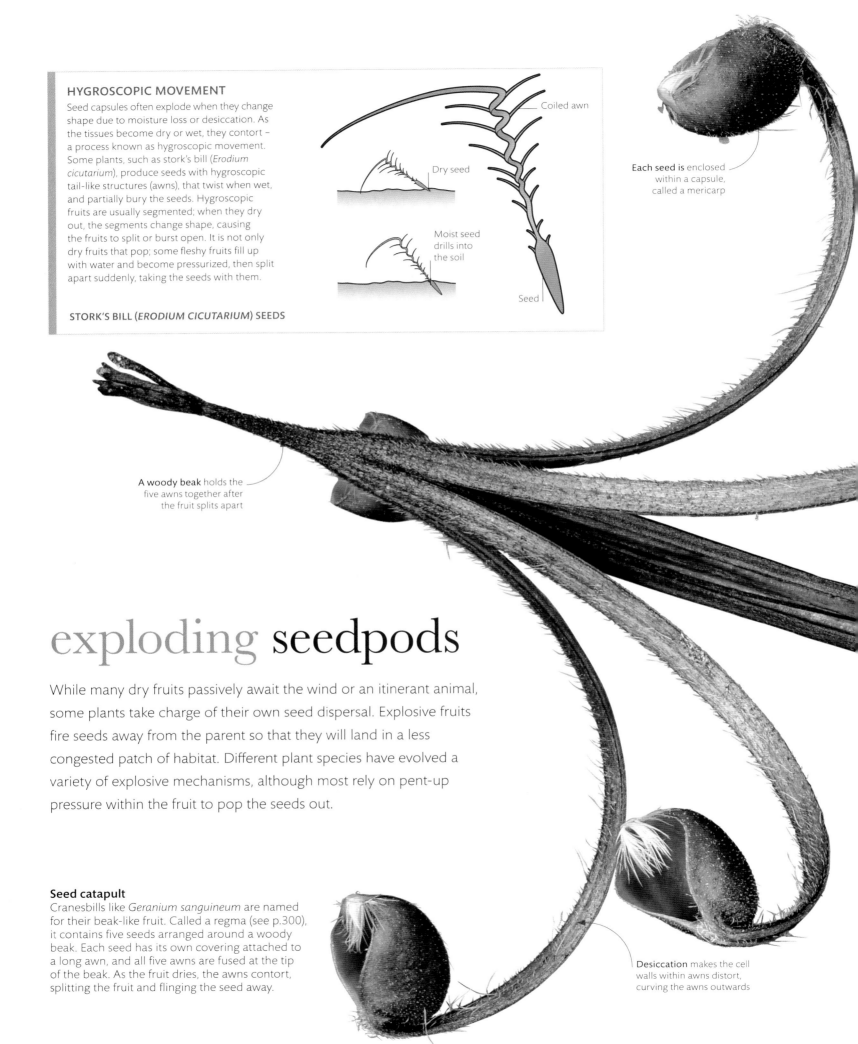

HYGROSCOPIC MOVEMENT

Seed capsules often explode when they change shape due to moisture loss or desiccation. As the tissues become dry or wet, they contort – a process known as hygroscopic movement. Some plants, such as stork's bill (*Erodium cicutarium*), produce seeds with hygroscopic tail-like structures (awns), that twist when wet, and partially bury the seeds. Hygroscopic fruits are usually segmented; when they dry out, the segments change shape, causing the fruits to split or burst open. It is not only dry fruits that pop; some fleshy fruits fill up with water and become pressurized, then split apart suddenly, taking the seeds with them.

STORK'S BILL (*ERODIUM CICUTARIUM*) SEEDS

Coiled awn

Dry seed

Moist seed
drills into
the soil

Seed

Each seed is enclosed
within a capsule,
called a mericarp

A woody beak holds the
five awns together after
the fruit splits apart

exploding seedpods

While many dry fruits passively await the wind or an itinerant animal, some plants take charge of their own seed dispersal. Explosive fruits fire seeds away from the parent so that they will land in a less congested patch of habitat. Different plant species have evolved a variety of explosive mechanisms, although most rely on pent-up pressure within the fruit to pop the seeds out.

Seed catapult

Cranesbills like *Geranium sanguineum* are named for their beak-like fruit. Called a regma (see p.300), it contains five seeds arranged around a woody beak. Each seed has its own covering attached to a long awn, and all five awns are fused at the tip of the beak. As the fruit dries, the awns contort, splitting the fruit and flinging the seed away.

Desiccation makes the cell
walls within awns distort,
curving the awns outwards

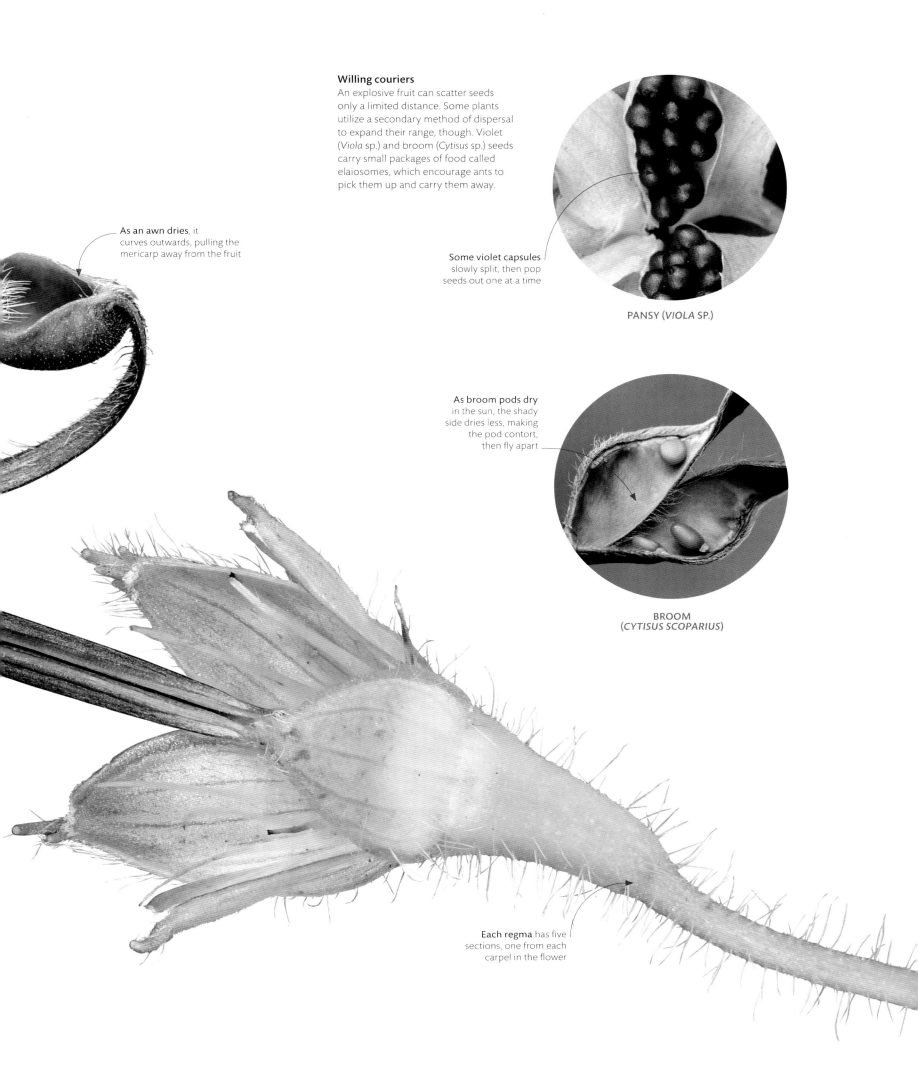

Willing couriers

An explosive fruit can scatter seeds only a limited distance. Some plants utilize a secondary method of dispersal to expand their range, though. Violet (*Viola* sp.) and broom (*Cytisus* sp.) seeds carry small packages of food called elaiosomes, which encourage ants to pick them up and carry them away.

As an awn dries, it curves outwards, pulling the mericarp away from the fruit

Some violet capsules slowly split, then pop seeds out one at a time

PANSY (*VIOLA* SP.)

As broom pods dry in the sun, the shady side dries less, making the pod contort, then fly apart

BROOM
(*CYTISUS SCOPARIUS*)

Each regma has five sections, one from each carpel in the flower

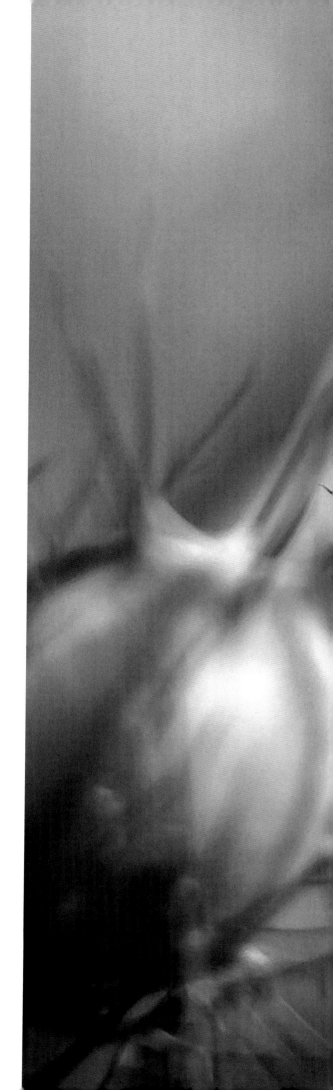

nigella

Nigella sativa, variously known as nigella, fennel flower, black cumin, black caraway, and Roman coriander, has an ancient association with human civilization, having been cultivated for around 3,600 years. The seeds are used as a spice sprinkled on bread and naan, or to produce a herbal oil.

Nigella's long history of cohabiting with humans makes unravelling its wild origins difficult. Some sources claim that the plant came from Mediterranean Europe, while others cite Asia or north Africa. Wild populations are still found in southern Turkey, Syria, and northern Iraq, so there is a strong possibility that it originated in the Middle East.

Standing up to 60 cm (2 ft) tall, nigella is a hardy annual that thrives in a variety of soil types. Despite the common names, it is not related to fennel (*Foeniculum vulgare*), cumin (*Cuminum cyminum*), caraway (*Carum carvi*), or coriander (*Coriandrum sativum*), all members of the carrot family. It is, in fact, a member of the buttercup family (Ranunculaceae) and a close relative of the popular ornamental love-in-a-mist (*Nigella damascena*). The delicate, beautiful blooms are most likely what first drew human attention to this plant. The identity of nigella's pollinators in the wild is not known, but bees perform the role elsewhere. After pollination, the plant's fruits swell into large, inflated capsules, each with several follicles containing numerous black seeds.

Irresistible to birds, *Nigella sativa*'s tiny, pungent seeds find favour with people too, being widely used as a spice in Indian and Middle Eastern cookery. In the ancient world the seeds and their oil were used to treat a wide range of ailments, and they are still used as a herbal remedy today.

Nigella sativa fruit
The pear-shaped seeds develop in a capsule with up to seven segments (follicles), each of which ends in a long projection formed from the style. The seeds are released when the fruit dries and the follicles rupture.

Numerous heavily dissected bracts, which are modified leaves, often support *Nigella* flowers

Style

Petals are more numerous in cultivated varieties; wild flowers have 5–10 petals

Cultivated flower
Love-in-a-mist is a popular garden annual, the flowers are often semi-double and deep blue in gardens (left), but paler and single in the wild.

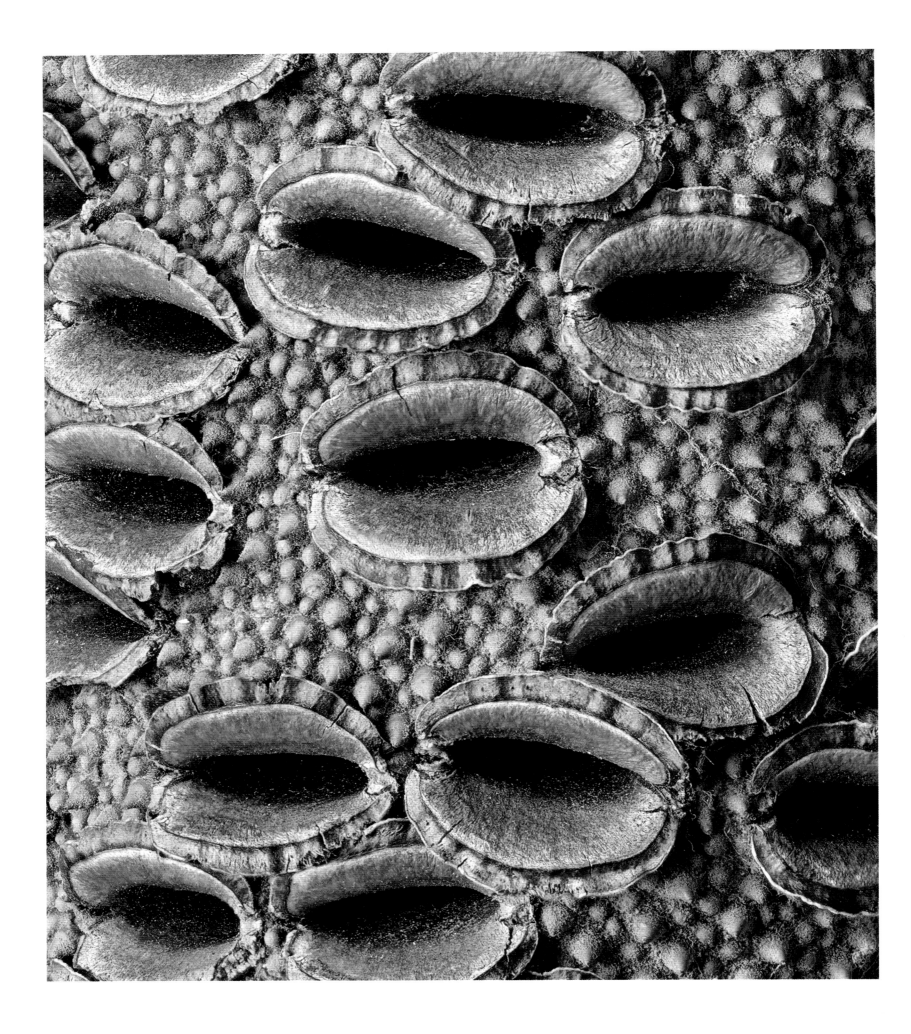

Heat treatment

The strange-looking, woody seedheads of some *Banksia* species retain seeds for years, and jettison them only after natural fires or artificial heat treatment makes their follicles snap open. Here the lip-like follicles have opened, releasing seeds.

seeds and fire

Ripe seeds usually separate spontaneously from parent plants. Some species, however, particularly those in harsh environments, only release their seed after an extreme environmental event. Fire is a common trigger for conifer forests, but it is also necessary for many other trees and shrubs, such as banksia in Australia. Although wildfires often kill younger trees, the heat makes their cone-like fruits open and drop their seeds, months or even years after they have matured.

AIDED BY FIRE

Banksia seeds need fire to trigger their release, but they also thrive in the environment that fire provides. Fire clears ground, eliminating competition from other plants, and fallen seeds sink easily into the soft ash left behind, which helps to protect them from the harsh sunlight.

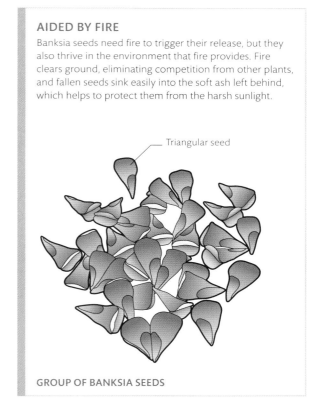

Triangular seed

GROUP OF BANKSIA SEEDS

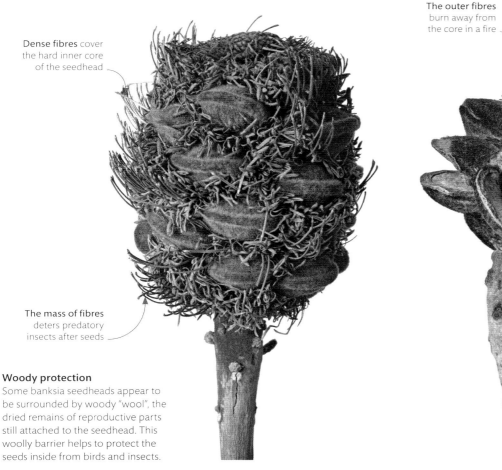

Dense fibres cover the hard inner core of the seedhead

The mass of fibres deters predatory insects after seeds

Woody protection
Some banksia seedheads appear to be surrounded by woody "wool", the dried remains of reproductive parts still attached to the seedhead. This woolly barrier helps to protect the seeds inside from birds and insects.

The outer fibres burn away from the core in a fire

Unopened follicles may indicate that no seed has set

The pedicel bows from the weight of the seedhead

Weighty release

Each flower of the water-dwelling sacred lotus (*Nelumbo nucifera*) produces an aggregate fruit, which matures into an extraordinary chambered seedhead that is roughly 7–12 cm (3–5 in) in diameter. As the numerous seeds mature, the seedhead shrinks. Eventually its stem bends under the weight, and the seeds drop out into the water.

Each seed, or nutlet, is around 1 cm (⅓ in) in diameter

The seed chamber enlarges as the drying seedhead shrinks

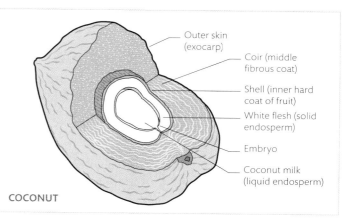

SEEDS DESIGNED TO FLOAT

Air pockets between the thick hairy fibres (coir) of a coconut keep it afloat in water. This husk is sandwiched between a protective outer skin (exocarp) and a hard inner shell (endocarp). The coconut "flesh" is food storage tissue (endosperm) that feeds the seed when it germinates, and the "milk" keeps it hydrated. Even after 4,800 km (3,000 miles) in seawater, a coconut can still produce seedlings.

Outer skin (exocarp)

Coir (middle fibrous coat)

Shell (inner hard coat of fruit)

White flesh (solid endosperm)

Embryo

Coconut milk (liquid endosperm)

COCONUT

water dispersal

Water transports many seeds – a process known as hydrochory. Obviously, wetland plants such as lotuses cast their seed into ponds, rivers, and streams, but the seeds of species such as harebells and silver birch may also be carried in this way. These short freshwater cruises pale in comparison, though, to the epic ocean voyages of tropical seeds such as coconuts.

Protective coating

Seeds need robust seed coats if they are to survive prolonged contact with water. Lotus seed coats are rock-hard, which makes them almost impermeable to water and helps to prevent the seeds from degrading.

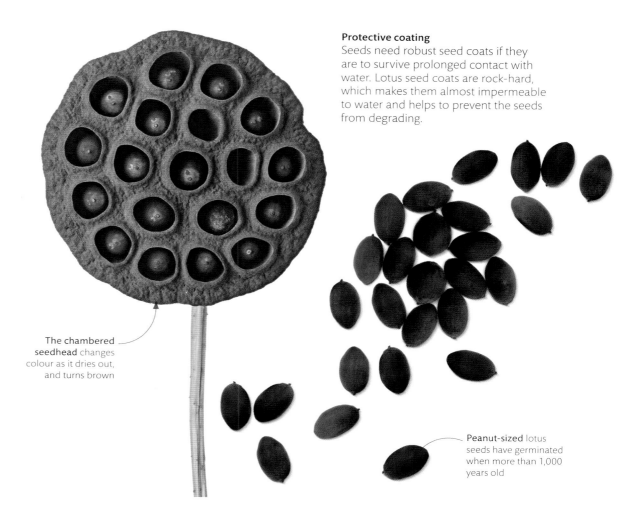

The chambered seedhead changes colour as it dries out, and turns brown

Peanut-sized lotus seeds have germinated when more than 1,000 years old

Exotic fruit
This chromolithograph shows the parts of the papaya plant (*Carica papaya*). It is one of the 40 plates on the flora of Java published in Berthe Hoola van Nooten's *Fleurs, Fruits et Feuillages de l'île de Java* (1863–64).

Exotic fruit
This chromolithograph shows the parts of the papaya plant (*Carica papaya*). It is one of the 40 plates on the flora of Java published in Berthe Hoola van Nooten's *Fleurs, Fruits et Feuillages de l'île de Java* (1863–64).

plants in art

painting the world

Nutmeg foliage, flowers, and fruit
North reveals the different stages of growth of a common culinary spice in this painting, made during her stay in Jamaica's Blue Mountains on her first major expedition in 1871–72. The flowers, foliage, and fruit of the nutmeg (*Myristica fragrans*) are shown with a butterfly (*Papilio polydamas*) and a hummingbird (*Mellisuga minima*).

The golden age of botany in the 18th and 19th centuries was characterized by the exploits of explorers and illustrators engaged in a global quest for specimens. Women were largely excluded from scientific discovery, but some intrepid individuals managed to set off on expeditions nonetheless, in search of new plants that they recorded in exquisite paintings.

Marianne North (1830–90) was a remarkable Victorian biologist and artist who, in 1871 at the age of 41, began to travel the world, recording flora in her paintings. Her gallery of 832 landscapes, plant, bird, and animal portraits at Kew Gardens in London offered the Victorian public an insight into the natural habitats of exotic specimens before the advent of colour photography. North's family connections and wealth made it possible for her to spend 13 years travelling across all continents, but the courage and energy were all her own.

At about the same age, Dutch-born Berthe Hoola van Nooten (1817–92), a penniless widow with an interest in botany, found herself in Batavia (Jakarta) on Java. Here she survived by selling her botanical paintings as chromolithographs; the Queen of the Netherlands supported her exquisite publication on Java's flora.

A century later, Margaret Mee (1909–88) began 30 years of study and painting in the Amazon rainforest. She recorded several new specimens, some of which were named after her, and painted plants against the rainforest backdrop.

Natural setting
The emerging fruits of the towering ackee (*Blighia sapida*) were painted by North in a natural setting in Jamaica. This West African plant was brought to Jamaica by Captain William Bligh and is named after him.

> " I had long dreamed of going to some tropical country to paint its peculiar vegetation on the spot in natural abundant luxuriance. "

MARIANNE NORTH, *RECOLLECTIONS OF A HAPPY LIFE*, 1892

Bulbils grow in the junction between leaf and rachis

A new plantlet forming from bulbil with leaves partially unrolled

Diplazium sori are arranged in a distinctive V-shape

Back-up plan
Diplazium proliferum can also reproduce from spores. These are held in structures called sori, situated along the veinlets on the underside of each leaflet.

natural clones

Some plants have evolved more than one way of creating new generations. All ferns, for example, reproduce by spores, but many also make clones of themselves by means of bulbils – tiny secondary bulbs that grow where leaves meet the main stem, or rachis, of a frond. The bulbils produce new plants when they drop off the parent fern or form roots where a frond droops and touches the soil. These new plants are clones of the parent.

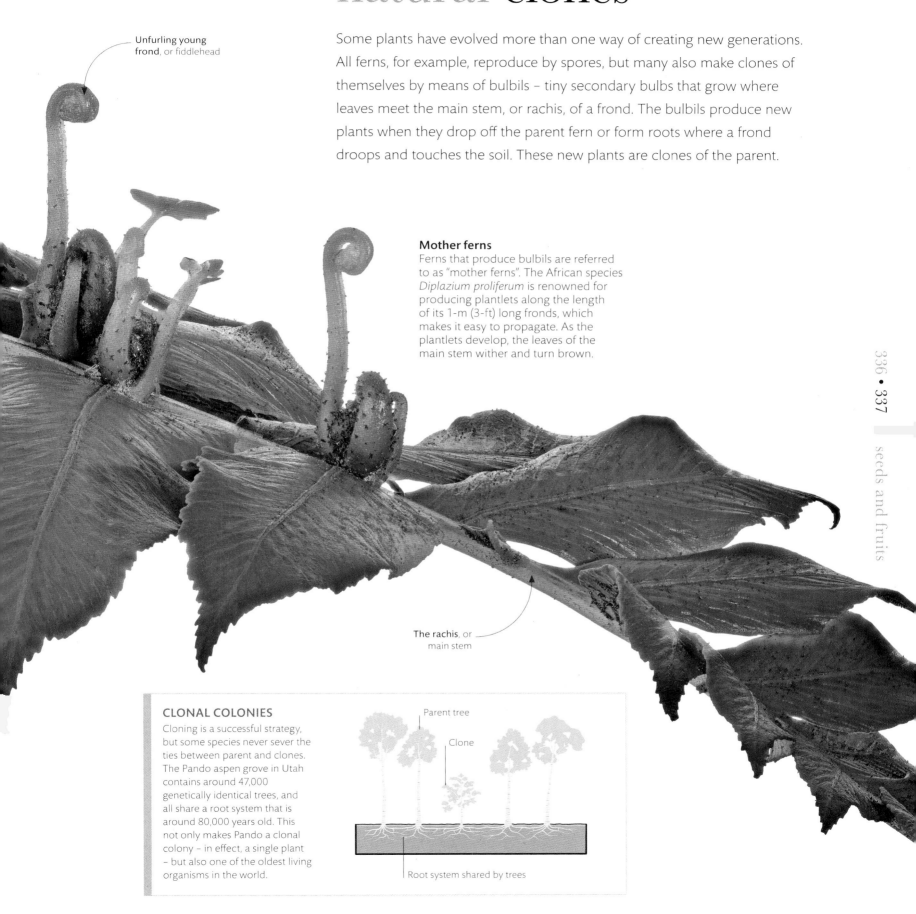

Unfurling young frond, or fiddlehead

Mother ferns
Ferns that produce bulbils are referred to as "mother ferns". The African species *Diplazium proliferum* is renowned for producing plantlets along the length of its 1-m (3-ft) long fronds, which makes it easy to propagate. As the plantlets develop, the leaves of the main stem wither and turn brown.

The rachis, or main stem

CLONAL COLONIES
Cloning is a successful strategy, but some species never sever the ties between parent and clones. The Pando aspen grove in Utah contains around 47,000 genetically identical trees, and all share a root system that is around 80,000 years old. This not only makes Pando a clonal colony – in effect, a single plant – but also one of the oldest living organisms in the world.

Parent tree

Clone

Root system shared by trees

The upper leaf surface does not produce sporangia

fern spores

Ferns do not flower – they produce spores in structures called sporangia (singular: sporangium) on the underside of their leaves. Sporangia are usually grouped in clusters, or sori, on their fronds, in patterns specific to each species. In some species, the immature sori are protected by a membrane known as the indusium. Each sorus contains many sporangia and thousands of spores, which are dispersed by the wind once mature.

seeds and fruits

Sori are arranged in lines on either side of the pronounced midrib

The round sori produced between veins sometimes fuse into lines

Sori are scattered across the entire frond, except over the major veins

ASPLENIUM AUSTRALASICUM

SELLIGUEA PLANTAGINEA

TECTARIA PICA

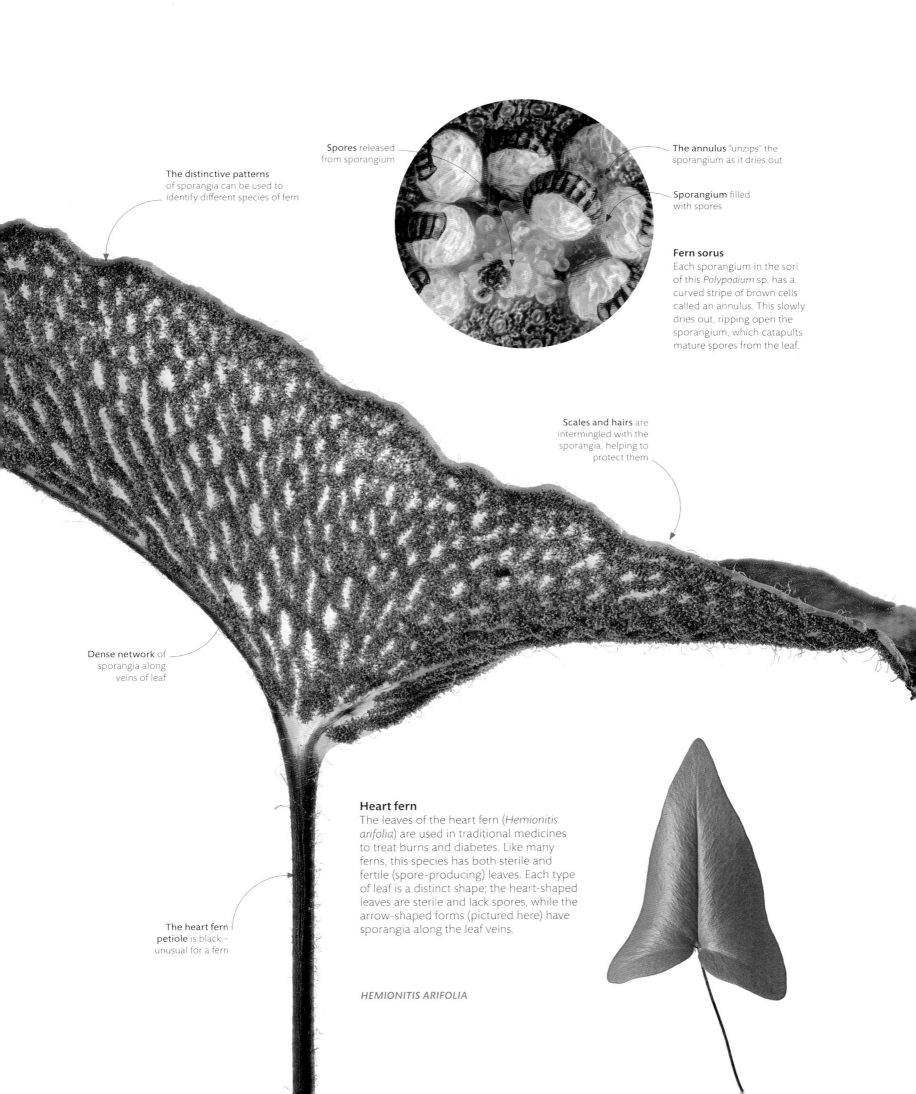

The distinctive patterns of sporangia can be used to identify different species of fern

Spores released from sporangium

The **annulus** "unzips" the sporangium as it dries out

Sporangium filled with spores

Fern sorus
Each sporangium in the sori of this *Polypodium* sp. has a curved stripe of brown cells called an annulus. This slowly dries out, ripping open the sporangium, which catapults mature spores from the leaf.

Scales and hairs are intermingled with the sporangia, helping to protect them

Dense network of sporangia along veins of leaf

Heart fern
The leaves of the heart fern (*Hemionitis arifolia*) are used in traditional medicines to treat burns and diabetes. Like many ferns, this species has both sterile and fertile (spore-producing) leaves. Each type of leaf is a distinct shape; the heart-shaped leaves are sterile and lack spores, while the arrow-shaped forms (pictured here) have sporangia along the leaf veins.

The heart fern petiole is black – unusual for a fern

HEMIONITIS ARIFOLIA

Cups of spores
The sporangia of *Lecanopteris carnosa* are formed in deep "cups" along the underside of the leaf. The cups fold back onto the upper surface of the leaf.

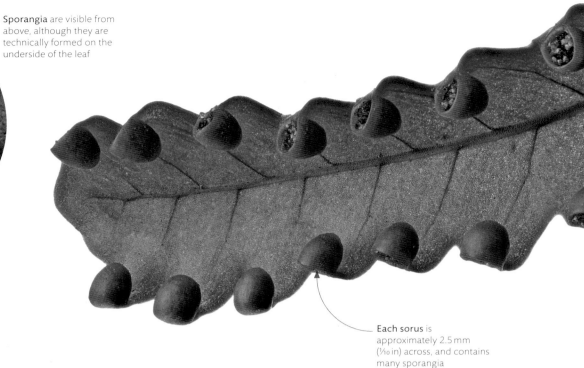

Each sorus is approximately 2.5 mm (1⁄10 in) across, and contains many sporangia

spore to fern

A wind-dispersed spore that lands in a suitable habitat will germinate into a tiny plantlet. Unlike a seed, the plantlet that grows from the spore – a gametophyte – has only one set of chromosomes. It makes gametes (sperm and eggs) that combine and grow into the more complex plant (sporophyte), which has two sets of chromosomes and is recognizable as a fern.

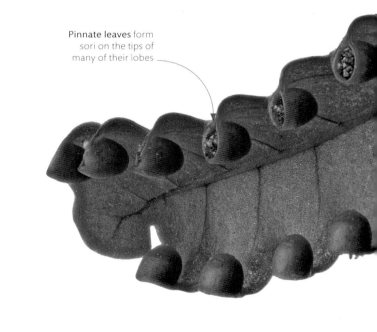

Pinnate leaves form sori on the tips of many of their lobes

ALTERNATION OF GENERATIONS

Fern gametophytes produce sperm, which travel through the water in their damp habitats to reach and fertilize the egg cell. The sporophyte develops from the fertilized egg (zygote). This alternation of generations, between gametophyte and sporophyte, occurs in flowering plants too, but flowering plant gametophytes (embryo sac and pollen grains) are very small and dependent on the sporophyte.

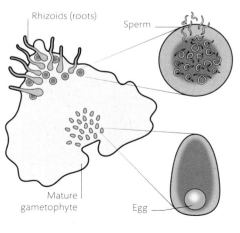

Rhizoids (roots)

Sperm

Mature gametophyte

Egg

The sporangia are held horizontally above the ground

The leaves reach up to 90 cm (3 ft) in length

Life in the canopy

The ant fern (*Lecanopteris carnosa*) grows on rainforest trees in Indonesia. It produces groups of four spores (tetrads) connected by filaments. These act like a parachute, helping to spread the spores to new trees by wind. Once germinated, the gametophytes are more likely to mate with each other than self-fertilize, improving the genetic diversity of the resulting sporophytes.

Spores may also be dispersed by ants

Tangled tetrads of spores will be dispersed by the wind

The cups holding sporangia do not usually photosynthesize

Pinnae (leaflets) near the tip of the frond form sori; those nearer the base are often sterile

Polytrichum sp.

haircap mosses

Seemingly insignificant and yet highly successful, tiny mosses have been around since at least the Permian Period, some 300 million years ago, and they can still be found growing on every continent today. Some of the most commonly encountered are the haircap mosses of the *Polytrichum* genus.

Haircap mosses have a two-phase life cycle known as alternation of generations. The gametophyte phase is the green, leafy part of the moss. The sporophyte phase is the hair-like growth that can sometimes be seen spouting from the moss stem — hence the common name of haircap. At the tip of the hair is a capsule holding spores. These two phases are genetically distinct. The gametophyte produces the sex cells (sperm and eggs) and has one set of chromosomes. Once the gametophyte is fertilized, the spore-bearing sporophyte, which has two sets of chromosomes,

grows out of the top of the moss. When dispersed spores land in a favourable spot, they grow into new gametophytes.

Mosses evolved long before plants developed vascular tissues for carrying water and nutrients, and most mosses must remain in contact with water for a portion of their lives to survive. However, thanks to a form of convergent evolution, haircap mosses possess rudimentary vascular tissues, allowing them to grow taller and survive longer periods of dehydration. As a result, haircap mosses have conquered many habitats that are too dry for most of their relatives.

Being so hardy, *Polytrichum* mosses play a key role in ecosystem regeneration, and are often the first to colonize barren soils. Patches of haircap moss prevent erosion, and their ability to retain water and reduce temperatures makes them good sites for other plants to germinate.

Hair-like sporophytes
Depending on humidity, the capsules open and close to release their spores when conditions are favourable for dispersal. The sporophyte is entirely dependent on the photosynthetic gametophyte for water and nutrients.

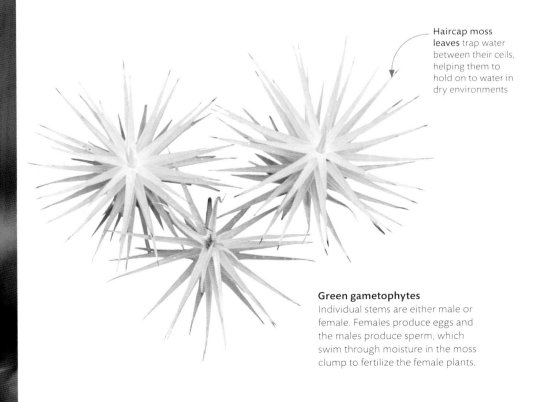

Haircap moss leaves trap water between their cells, helping them to hold on to water in dry environments

Green gametophytes
Individual stems are either male or female. Females produce eggs and the males produce sperm, which swim through moisture in the moss clump to fertilize the female plants.

glossary

ABAXIAL The side of an organ that faces away from the stem or supporting structure; typically used to describe the lower surface of a leaf.

ACCESSORY FRUIT A fruit consisting of the ovary together with another plant part, such as the swollen end of a flower stalk. Examples include apples and rose hips. Also known as false fruit.

ACHENE A dry fruit that does not open and contains a single seed.

ADVENTITIOUS Arising from places where growths do not normally occur: adventitious roots may arise from stems, for example.

AERIAL ROOT A root that grows from the stem of a plant that is located above the ground.

AGGREGATE FRUIT A compound fruit that develops from several ovaries. The ovaries are all from the carpels of a single flower and the separate fruits are joined together. Examples are raspberries and blackberries.

ALGAE A group of simple, flowerless, mainly aquatic, plant-like organisms that contain the green pigment chlorophyll but do not have true stems, roots, leaves, and vascular tissue. An example is seaweed.

ANGIOSPERM A flowering plant that bears ovules, later seeds, enclosed in ovaries. Angiosperms are divided into two main groups for classification: monocots and eudicots.

ANNUAL A plant that completes its entire life cycle – germination, flowering, seeding, and dying – in one growing season.

ANNULUS (PL. ANNULI) The ring of thick-walled cells involved in opening a fern's sporangium and releasing the spores.

ANTHER The part of a flower's stamen that produces pollen; it is usually borne on a filament.

ANTHOCYANIN Plant pigment molecules that are responsible for red, blue, and purple colours in leaves and flowers.

APEX The tip or growing point of a leaf, shoot, or root.

APOMIXIS The process of asexual reproduction.

ARIL A berry-like, fleshy, hairy, or spongy layer around some seeds.

AWN A bristle that grows from the spikelets of certain grasses, including cultivated cereals.

AXIL The upper angle between a stem and a leaf, where an axillary bud develops.

AXILLARY BUD A bud that develops in the axil of a leaf.

BALAUSTA A fruit with a tough skin, or pericarp, and many cells, each containing a seed. A typical example is a pomegranate.

BARK The tough covering on woody roots, trunks, and branches.

BERRY A fruit with soft, juicy flesh surrounding one or more seeds that have developed from a single ovary.

BIENNIAL A plant that flowers and dies in the second growing season after germination.

BIPINNATE A compound leaf whose leaflets are divided into yet smaller leaflets, such as mimosa leaves.

BISEXUAL *See* Perfect flower.

BLADE The whole part of a leaf except for its stalk (petiole). The shape of a leaf blade and its edges – or margins – are important characteristics of a plant.

BRACT A leaf that has modified into an attractive or protective structure – usually to protect buds – around the base of a flower or flower cluster. Some bracts are large, brightly coloured, and resemble flower petals to attract beneficial insects, while others look like leaves, although they may be smaller and shaped differently from the other leaves on the plant.

BRACTEOLE Smaller version of a bract, growing from the flower stem above it.

BROADLEAVED Describes trees and shrubs that have broad, flat, usually deciduous leaves, in contrast to the narrow, needle-like leaves of conifers.

BROMELIAD A rosette-forming plant in the pineapple family (Bromeliaceae) that may also be an epiphyte. It has spirally arranged and sometimes toothed leaves, and stems that can become woody.

BUD An immature organ or shoot enclosing an embryonic branch, leaf, inflorescence, or flower.

BULB A modified underground bud that acts as a storage organ. It consists of one or more buds, and layers of swollen, colourless, fleshy scale leaves, packed with stored food, on a shortened, disc-like stem.

BULBIL A small bulb-like organ, often borne in a leaf axil, occasionally on a stem or flower head.

BUNDLE SHEATH Cylinder of cells surrounding a vascular bundle inside a plant leaf.

BURR A prickly or spiny dry fruit.

CALYX (PL. CALYCES) The outer part of a flower, formed from a ring of sepals, that is sometimes showy and brightly coloured but usually small and green. The calyx forms a cover that encloses the petals while in bud.

CAMBIUM A layer of tissue capable of producing the new cells that increase the girth of stems and roots.

CAPITULUM (PL. CAPITULA) A group of flowers (inflorescence) on a stem that together look like a single flower head. An example is a sunflower.

CAPSULE A dry fruit, containing many seeds, that has developed from an ovary formed from two or more carpels. It splits open when ripe to release its seeds.

CAROTENOIDS Plant pigment molecules responsible for yellow and orange hues.

CARPEL The female reproductive part of a flower that consists of an ovary, stigma, and style.

CARYOPSIS (PL. CARYOPSES) A dry, single-seeded fruit that is indehiscent. Grasses typically have rows or clusters of edible caryopses, or grains.

CATKIN A long, thin cluster of small flowers with inconspicuous or no petals that hangs from a tree such as a hazel or birch.

CAULIFLOROUS A term used to describe flowers and fruits that develop directly on a tree's trunk or branches rather than at the ends of its twigs.

CHASMOGAMOUS Having open blooming flowers with exposed reproductive organs for cross-pollination.

CHLOROPHYLL The green pigment inside plant cells that allows leaves and sometimes stems to absorb light and carry out photosynthesis.

CHLOROPLAST The particles inside plant cells that contain chlorophyll, where starch is formed during photosynthesis.

CIRCINATE Coiled inwards, as in fern fiddleheads.

CLADODE A modified stem that both resembles and performs the function of a leaf.

CLASS In taxonomy, the rank below kingdom and above order – for example, monocots and eudicots.

CLEISTOGAMOUS A flower that pollinates itself without the bloom opening. Opposite: Chasmogamous.

COLEOPTILE A sheath that protects the shoot emerging from a monocot seed as it grows through the soil.

COMPOUND LEAF A leaf composed of two or more similar parts (leaflets).

CONE The densely clustered bracts of conifers and some flowering plants, that often develop into a woody, seed-bearing structure, such as a pine cone.

CONIFERS Mostly evergreen trees or shrubs, that usually have needle-like leaves and naked seeds that develop on scales inside cones.

CORM A bulb-like underground swollen stem or stem base, often surrounded by a papery tunic.

COROLLA A ring of petals on a flower head.

CORTEX The region of tissue between the epidermis or bark and the vascular cylinder.

CORYMB A broad, flat-topped, or domed inflorescence of stalked flowers or flower heads arising at different levels on alternate sides of an axis.

COTYLEDON A seed leaf that acts as a food store or unfurls shortly after germination to fuel a seed's growth.

CREMOCARP A small, dry fruit that forms in two flattened halves, each containing one seed.

CROSS-FERTILIZATION The fertilization of the ovules of a flower as a result of cross-pollination.

CROSS-POLLINATION The transfer of pollen from the anthers of a flower on one plant to the stigma of a bloom on another plant. *See also* Self-pollination.

CUCURBIT A plant from the cucumber or gourd family (Cucurbitales) which includes melon, pumpkin, and squash.

CULM The jointed, usually hollow, flowering stem of a grass or bamboo.

CULTIVAR (CV.) A contraction of "cultivated variety", used to describe a plant that usually only exists in human cultivation.

CUPULE A cup-shaped structure made of bracts joined together.

CUTICLE A protective, waxy, water-repellent coating of the outer cells of the eperdermis of some plants.

CYME A flat or round-topped, branched inflorescence with each axis ending in a flower, the oldest at the centre and the youngest arising in succession from the axils of bracteoles (secondary bracts).

DECIDUOUS Used to describe plants that shed their leaves at the end of the growing season and then renew them at the beginning of the next. Semi-deciduous plants lose only some of their leaves at the end of the growing season.

DEHISCENT FRUIT A dry fruit that splits or bursts to release its seeds.

DICHASIUM *See* Inflorescence.

DICOT, DICOTYLEDON A term – now considered obsolete or incorrect in evolutionary history – that was used to describe a flowering plant with two seed leaves, or cotyledons. *See also* Eudicot, Monocot.

DIPLOID Having two sets of chromosomes; common in most plant tissue cells. *See also* Haploid.

DIOECIOUS A plant that bears unisexual flowers, with male and female blooms occuring on separate plants. *See also* Monoecious.

DIVISION The propagation of a plant by splitting it into two or more parts, each with its own section of root system and one or more shoots or dormant buds.

DOMATIUM (PL. DOMATIUM) A structure produced by a plant that is inhabited by animals, usually cavities in the roots, stems, or leaf veins, and often associated with ants.

DRIP TIP The edge of a leaf or leaflet that helps to direct rainwater run-off.

DRUPE A fleshy fruit containing a seed with a hard coat (endocarp).

EMERGENT Coming out of, or emerging from.

ENDOCARP The innermost layer of the pericarp of a fruit.

EPIDERMIS The outer protective layer of cells of a plant.

EPIPHYTE A plant that grows on the surface of another plant without being parasitic or stealing nutrients from its host; it obtains moisture and nutrients from the atmosphere without rooting into the soil.

EUDICOT, EUDICOTYLEDON A flowering plant that has two seed leaves, or cotyledons, including many plants formerly described as "dicots". Most eudicots have broad leaves with branching veins and the floral parts, such as petals and sepals, arranged in groups of multiples of four or five. *See also* Monocot.

EVERGREEN Describes plants that retain their foliage for more than one growing season; semi-evergreen plants retain only a small portion of their leaves for more than one season.

EXOCARP The outer layer of the pericarp of a fruit. The exocarp is often thin and hard, or like a skin.

EXODERMIS A specialized layer in a root beneath the epidermis or velamen.

FALSE FRUIT *See* Accessory fruit.

FAMILY In plant classification, a group of related genera; the family Rosaceae, for example, includes the genera *Rosa*, *Sorbus*, *Rubus*, *Prunus*, and *Pyracantha*.

FERN A flowerless, spore-producing plant consisting of roots, stems, and leaf-like fronds. *See also* Frond.

FIDDLEHEAD The coiled young leaf of a fern.

FILAMENT The stalk that bears the anther in a flower.

FLORET A small flower, usually one of many florets that make up a composite flower such as a daisy.

FLOWER The reproductive organ of a great many plant genera. Each flower consists of an axis which bears four types of reproductive organs: sepals, petals, stamens, and carpels.

FOLLICLE A dry fruit, similar to a pod, that develops from a single-chambered ovary with one seam that splits open to release the seeds. Most follicles are aggregate fruits.

FROND 1. The leaf-like organ of a fern. Some ferns produce both barren fronds and fertile fronds, which bear spores. 2. Large, usually compound leaves such as those of palms.

FRUIT The fertilized, ripe ovary of a plant, which contains one or more seeds, such as berries, hips, capsules, or nuts. The term is also used to describe edible fruits.

FRUITLET A small fruit that comprises part of an aggregate fruit such as blackberry.

FUNGUS A single-celled or multi-celled organism that is a member of the separate kingdom Fungi: moulds, yeasts, and mushrooms, for example.

FUNICLE The tiny stalk inside a pod from which a seed hangs.

GENUS (PL. GENERA) A category in plant classification ranked between family and species.

GERMINATION The physical and chemical changes that take place when a seed starts to grow and develop into a plant.

GLABROUS Smooth and hairless.

GLUME A scaly, protective bract, usually one of a pair, at the base of a grass or sedge spikelet.

GYMNOSPERM A plant with seeds that develop without an ovary to enclose and protect them while they mature. Most gymnosperms are conifers, whose seeds form on scales and mature within cones.

HARDY Describes plants that can withstand freezing temperatures in winter.

HAPLOID A cell with only one set of chromosomes. *See also* Diploid.

HAUSTORIA Specialized parasitic plant roots that penetrate the tissues of a host plant.

HEMIPARASITE A parasite that has green leaves and can photosynthesize. An example is mistletoe (*Viscum album*).

HERBACEOUS A non-woody plant in which the upper parts die down to a rootstock at the end of the growing season. The term is mainly used to describe perennial plants, although botanically it also applies to annuals and biennials.

HERMAPHRODITE Plant species with flowers in which the male stamens and female pistil are present together in single bisexual or perfect flowers.

HESPERIDIUM The fruit of a citrus plant with a thick, leathery rind, such as a lemon or orange.

HETEROPHYLLOUS Plants with leaves of different shapes or forms on the same plant that may be adapted to perform in particular conditions such as sun or shade.

HIP A cup-shaped receptacle containing dry, one-seeded fruits that are covered with small hairs.

HOLOPARASITE A parasitic plant with no leaves that is totally dependent on its host for food and water.

HYBRID The offspring of genetically different parent plants. Hybrids between species of the same genus are described as interspecific hybrids; those between different – but usually closely related – genera are known as intergeneric hybrids.

HYSTERANTHOUS A plant whose flowers open before its leaves, such as forsythia or witch hazel.

IMPERFECT FLOWER Flowers that contain either only male or only female reproductive organs. Also known as unisexual.

INDEHISCENT Describes a fruit that does not split open to release its seeds, such as a hazelnut.

INDUSIUM A thin flap of tissue that covers a fern's sorus.

INFLORESCENCE A group of flowers borne on a single axis (stem), such as a raceme, panicle, or cyme.

INTERNODE The portion of stem between two nodes.

INVOLUCRE A ring of leaf-like bracts below a flower head.

KEEL 1. A prominent longitudinal ridge, usually on the underside of a leaf, that resembles the keel of a boat. 2. The two lower, fused petals of a pea-like flower.

KINGDOM Any of the four main categories of living organisms, such as the plant kingdom.

LABELLUM A lip, particularly the prominent third petal of iris or orchid flowers. *See also* Lip.

LAMINA A broad, flat structure, for example, the blade of a leaf.

LATERAL A side growth that arises from a shoot or root.

LEAF Typically a thin, flat lamina (blade) growing out of a stem that is supported by a network of veins. Its main function is to collect the energy from sunlight that the plant needs in order to photosynthesize.

LEAFLET One of the subdivisions of a compound leaf.

LEGUME A dehiscent pod that splits along two sides to disperse ripe seed.

LEMMA The outermost of two bracts enclosing a grass flower. *See also* Palea.

LENTICEL A hole in the stem that allows gases to pass between the plant's cells and the air around them.

LIGNIN A hard substance in all vascular plants that enables them to grow upright and remain standing.

LINEAR Very narrow leaves that have parallel sides.

LIP A prominent lower lobe on a flower, formed by one or more fused petals or sepals. *See also* Labellum.

LIVERWORT A simple, flowerless plant that lacks true roots. It has leaf-like stems or lobed leaves, reproduces by shedding spores, and is usually found in damp habitats.

LOBED Describes a part of a plant, such as a leaf, that has curved or rounded parts.

LOCULE A compartment or chamber of an ovary or anther.

MARGIN The outer edge of a leaf.

MERISTEM Plant tissue that is able to divide to produce new cells. Both shoot or root tips can contain meristematic tissue and may be used for micropropagation.

MESOCARP The middle layer of the pericarp. In many fruits the mesocarp is the fleshy part of the fruit. In some pericarps the mesocarp is missing.

MESOPHYLL The soft, inner tissue (parenchyma) of a leaf, between the upper and lower layers of the epidermis, that contains chloroplasts for photosynthesis.

MICROPROPAGATION The growing of tiny bits of plant tissue in laboratories in order to produce new plants.

MIDRIB The primary, usually central, vein of a leaf.

MONOCARPIC A plant that flowers and fruits only once before it dies; such plants may take a number of years to reach flowering size.

MONOCOT, MONOCOTYLEDON A flowering plant that has only one cotyledon, or seed leaf, in the seed; it is also characterized by narrow, parallel-veined leaves. Monocot examples include lilies, irises, and grasses. *See also* Eudicot.

MONOECIOUS A plant with separate male and female flowers that are borne on the same plant. *See also* Dioecious.

MOSS A small, green, flowerless plant that has no true roots and grows in damp habitats. It reproduces by shedding spores.

MUCILAGE A gummy secretion present in various parts of plants, especially leaves.

MULTIPLE FRUIT A fruit that develops from several close-set flowers that fuse together to form a single fruit such as pineapple.

MYCORRHIZA A mutually beneficial (symbiotic) relationship between a fungus and the roots of a plant.

NECTAR A sweet, sugary substance secreted by a nectary, to attract insects and other pollinators to a flower.

NECTARY A gland that secretes nectar. Nectaries are most frequently located in the flower of a plant, but are sometimes found on the leaves or stems.

NODE A point of a stem from which one or more leaves, shoots, branches, or flowers arise.

NODULE 1. A small knob on a root that contains nitrogen-fixing bacteria.

2. Small swellings on a leaf (on the petiole, midrib, lamina, or margin) that contain bacteria.

NUT A one-seeded, indehiscent fruit with a tough or woody coat, for example, an acorn. Less specifically, all fruits and seeds with woody or leathery coats.

NUTRIENTS Minerals (mineral ions) used to develop proteins and other compounds required for plant growth.

OFFSET A small plant that develops from a shoot growing out of an axillary bud on the parent plant.

ORDER In taxonomy, the rank below class and above family.

OVARY The lower part of the carpel of a flower, containing one or more ovules; it may develop into a fruit after fertilization.

OVULE The part of an ovary that develops into the seed after pollination and fertilization.

OVULIFEROUS SCALES The scales of a female cone which bear ovules that become seeds once they have been fertilized.

PALEA (PL. PALEAE) The innermost of the two bracts enclosing a grass flower. *See also* Lemma.

PALMATE Has lobed leaflets that arise from a single point.

PANICLE A branched raceme.

PAPPUS An appendage, or tuft of appendages, that crowns the ovary or fruit in various seed plants and functions in wind dispersal of the fruit.

PARENCHYMA Soft plant tissue consisting of cells with thin walls.

PEDICEL The stalk bearing a single flower in an inflorescence.

PEDUNCLE The main stalk of an inflorescence, which holds a group of pedicels.

PENDENT Hanging downwards.

PEPO A many-seeded, hard-skinned berry that forms the fruit of cucurbits such as pumpkins, watermelons, and cucumbers.

PERENNIAL A plant that lives for more than two years.

PERFECT FLOWER Flowers that contain both male and female reproductive organs. Also known as bisexual or hermaphrodite.

PERFOLIATE Stalkless leaves or bracts that encircle the stem, so that the stem appears to pass through the leaf blade.

PERIANTH The collective term for the calyx and corolla, particularly when they are very similar in form, as in many bulb flowers.

PERICARP The wall of a fruit that develops from the maturing ovary wall. In fleshy fruits the pericarp often has three distinct layers: exocarp, mesocarp, and endocarp. The pericarp of dry fruits is papery or feathery, but on fleshy fruits it is succulent and soft.

PETAL A modified leaf, usually brightly coloured and sometimes scented, that attracts pollinators. A ring of petals on a flower is known as a corolla.

PETIOLE The stalk of a leaf.

PHLOEM The vascular tissue in plants that conducts sap containing nutrients (produced by photosynthesis) from the leaves to other parts of a plant.

PHOTOSYNTHESIS The process by which the energy in sunlight is trapped by green plants and used to carry out

a chain of chemical reactions to create nutrients from carbon dioxide and water. A by-product is oxygen.

PHYLLOTAXIS The arrangement of leaves on a stem or branch.

PHYTOTELMA (PL. PHYTOTELMATA) A water-filled cavity in a plant that serves as a habitat.

PIN FLOWER A flower with a long style and relatively short stamens. Opposite: Thrum flower.

PINNATE The arrangement of leaflets on opposite sides of the central stalk of a compound leaf.

PIONEER SPECIES A species that colonizes a new environment, for example after volcanic eruptions or fire, and starts a plant succession.

PISTIL *See* Carpel.

PLANT Living organism, ranging from trees to grasses and flowers, that produces its own food by photosynthesis.

PLANTLET A young plant that develops on the leaf of a parent plant.

PLUMULE The first shoot that emerges from a seed when it germinates.

PNEUMATOPHORE An erect aerial root that protrudes upwards through swampy soil with the ability to exchange gases or "breathe". Often found in mangroves.

POD A flattened, dry fruit that develops from a single ovary with one chamber.

POLLEN Small grains, formed in the anther of seed-bearing plants, which contain the male reproductive cells of the flower.

POLLINATION The transfer of pollen from an anther to the stigma of a flower.

POLLINATOR 1. The agent or means by which pollination is carried out, for example, via insects, birds, wind. 2. A plant required to ensure that seeds set on another self- or partially self-sterile plant.

POME The fleshy part of an apple or related fruit, consisting of an enlarged receptacle and the ovary and seeds.

PRICKLE A sharp outgrowth from the epidermis or cortex of a plant, which can be detached without tearing the part of the plant from which it is growing.

PROPAGATE To increase plants by seed or by vegetative means.

PROTANDROUS Hermaphroditic plants that are functionally male before becoming functionally female. Opposite: Protogynous.

PROTOGYNOUS Hermaphroditic plants that are functionally female before becoming functionally male. Opposite: Protandrous.

PSEUDOBULB A thickened, bulb-like stem arising from a (sometimes very short) rhizome.

PTERIDOPHYTE A fern or fern-ally, such as horsetail or clubmoss, that reproduces using alternating generations, with the main generation producing spores.

RACEME A cluster of several or many separate flower heads borne singly on short stalks along a central stem, with the youngest flowers at the tip.

RACHIS The main axis of a compound leaf or inflorescence.

RADICLE The root of a plant embryo. The radicle is normally the first organ to appear when a seed germinates.

RECEPTACLE The enlarged or elongated tip of the stem from which all parts of a simple flower arise.

RECURVED Arched backwards.

REFLEXED Bent completely backwards.

REGMA (PL. REGMATA) A type of dry fruit consisting of three or more fused carpels that break apart explosively when mature.

REPLUM A thin partition between the parts in certain fruits, such as some legumes.

RESIN A thick, sticky substance formed of organic compounds that a tree produces to heal wounds in its bark that have been inflicted by pests or caused by physical damage.

RHIZOME A creeping underground stem that acts as a storage organ and produces shoots at its apex and along its length.

RHIZOSPHERE A root system and the immediate surrounding substrate.

ROOT The part of a plant, normally underground, that anchors it in the soil and through which water and nutrients are absorbed.

ROOT CAP A hood-shaped cap at the root tip that continually produces new cells to protect the root from abrasion as it grows through the soil.

ROOT HAIR A thread-like growth that develops behind the root cap. Root hairs extend the surface area of a root and increase the amount of water and nutrients it can absorb.

ROSETTE A cluster of leaves radiating from approximately the same point, often at ground level at the base of a very short stem.

RUNNER A horizontally spreading, usually slender, stem that runs above ground and roots at the nodes to form new plants. *See also* Stolon.

SAMARA A dry, indehiscent, one-seeded fruit that has "wings" to ensure dispersal by the wind. Examples are ash, maple, and sycamore.

SAP The juice of a plant contained in the cells and vascular tissue.

SCALE A reduced leaf, usually membranous, that covers and protects buds, bulbs, and catkins.

SCHIZOCARP A papery, dry fruit, which breaks up into enclosed, single-seeded units that disperse separately when the seeds are ripe.

SCLEREID A plant cell with lignified, pitted walls.

SEAM (OR SUTURE) The edge of a pod where it breaks open.

SEED The ripened, fertilized ovule, containing a dormant embryo capable of developing into an adult plant.

SEEDLING A young plant that has developed from a seed.

SELF-INCOMPATIBLE Describes a plant that is unable to produce viable seed by fertilizing itself and needs a different pollinator in order for fertilization to take place. Also known as self-sterile.

SELF-POLLINATION The transfer of pollen from the anthers to the stigma of the same flower, or alternatively to another flower on the same plant. *See also* Cross-pollination.

SELF-STERILE *See* Self-incompatible.

SEPAL The outer whorl of the perianth of a flower, usually small and green, but sometimes coloured and petal-like.

SEPTUM A partition that separates two chambers in a fruit.

SHOOT A developed bud or young stem.

SIMPLE FRUIT A fruit that forms from a single ovary. Examples include berries, drupes, and nuts.

SIMPLE LEAF A leaf that is formed in one piece.

SORUS (PL. SORI) 1. A cluster of sporangia on the underside of a fern leaf. 2. A spore-producing structure of some lichens and fungi.

SPADIX A fleshy flower spike that bears numerous small flowers, usually sheathed by a spathe.

SPATHE A bract that surrounds a single flower or a spadix.

SPECIES (SP.) In plant classification, a class of plants whose members have the same main characteristics and are able to breed with one another.

SPIKE A long flower cluster with individual flowers borne on very short stalks or attached directly to the main stem.

SPIKELET Group of florets of a grass, enclosed by protective glumes.

SPINE A stiff, sharp-tipped, modified leaf or leaf part, such as a stipule or petiole.

SPORE The minute, reproductive structure of flowerless plants, such as ferns, fungi, and mosses.

SPORANGIUM (PL. SPORANGIA) A body that produces spores on a fern.

SPUR 1. A hollow projection from a petal, which often produces nectar. 2. A short branch that bears a group of flower buds (such as those found on fruit trees).

STAMEN The male reproductive part of a flower comprising the pollen-producing anther and usually its supporting filament or stalk.

STAMINOID A sterile structure resembling a stamen.

STANDARD The upper petal in certain flowers in the pea family.

STEM The main axis of a plant, usually above ground, that supports structures such as branches, leaves, flowers, and fruit.

STIGMA The female part of a flower that receives pollen before fertilization. The stigma is situated at the tip of the style.

STIPULE A leafy outgrowth, often one of a pair.

STOLON A horizontally spreading or arching stem, usually above ground, that roots at its tip to produce a new plant. Often confused with a runner.

STOMA (PL. STOMATA) A microscopic pore in the surface of aerial parts of plants (leaves and stems), allowing transpiration to take place.

STRIATE Striped

STYLE The stalk that connects the stigma to the ovary in flowers.

SUBMERGENT A plant that lives entirely underwater.

SUBSPECIES A major division of a species, where the distinction is not complete.

SUCCULENT A drought-resistant plant with thick, fleshy leaves or stems adapted to store water. All cacti are succulents.

SUCKER A new shoot that develops from the roots or the base of a plant and rises from below ground level.

SYMBIOSIS Living together in a mutually beneficial relationship.

SYMBIOTIC Mutually beneficial.

TAP ROOT The primary, downward-growing root of a plant such as dandelion.

TENDRIL A modified leaf, branch, or stem, usually long and slender, that can attach itself to a support.

TEPAL A single segment of a perianth that cannot be distinguished as either a sepal or a petal, as in crocuses or lilies.

TERMINAL BUD A bud that forms at the apex or tip of a stem.

TESTA The hard, protective coating around a fertilized seed that prevents water from entering the seed until it is ready to germinate.

THIGMOTROPISM A plant's ability to grow, bend, and twine in response to a touch stimulus.

THORN A modified stipule or simple outgrowth from a stem that forms a sharp, pointed end.

THRUM FLOWER A flower with a short style in which only the stamens are visible in the throat of the corolla. Opposite: Pin flower.

THYRSE A compound inflorescence with numerous flowering stalks that branch in pairs from the main stem.

TRANSPIRATION The loss of water by evaporation from leaves and stems.

TRICHOME Any type of outgrowth from the surface tissue of a plant, such as a hair, scale, or prickle.

TRIFOLIOLATE Describes a compound leaf that has three leaflets growing from the same point.

TUBER A swollen, usually underground, organ derived from a stem or a root, used for food storage.

UMBEL A flat or round-topped inflorescence in which the flower stalks grow from a single point at the top of a supporting stem.

UNISEXUAL Flower producing either pollen (male) or ovules (female).

VARIEGATED Describes irregular arrangements of pigments, usually the result of either mutuation or disease, mainly in leaves.

VARIETY Another division of a species, generally differing from the type in one character only.

VASCULAR BUNDLE A unit of water-conducting xylem and food-conducting phloem tissue grouped together in the vein or stem of a plant leaf.

VASCULAR CYLINDER The central cord of vascular tissue.

VASCULAR PLANT A plant that has food-conducting tissues (the phloem) and water-conducting tissues (xylem).

VEIN A vascular structure in a leaf, surrounded by the bundle sheath, often visible as lines on a leaf surface.

VELAMEN Water-absorbing tissue that covers the aerial roots of certain plants, including many epiphytes.

VENATION The arrangement of veins in a leaf.

VERNATION The folding of leaves in a bud.

VIVIPAROUS 1. Describes a plant that forms plantlets on leaves, flower heads, or stems. 2. Also applied loosely to plants that produce bulbils on bulbs.

WHORL An arrangement of three or more organs that all arise from the same point.

WINGED FRUIT A fruit with fine, papery structures that are shaped like wings to help carry the fruits through the air. *See also* Samara.

XYLEM The woody part of plants, consisting of supporting and water-conducting vascular tissue.

ZYGOMORPHIC Describes a flower that can only be cut in one plane to produce two halves that are mirror images of each other.

index

Page numbers in **bold** relate to pages with the most information. Page numbers in *italics* relate to information found in diagrams.

A

Abutilon sp. 198
Acacia sp. 79, 88, 160
    A. karroo (sweet thorn) **158–59**
Acanthus, Morris 116
accessory fruits *295*
Acer sp. 66, 67
    A. japonicum 131
    A. palmatum (Japanese maple) 136, **152–53**
    A. platanoides 'Drummondii' 150
    A. pseudoplatanus (sycamore) 118–19, 314
    A. saccharum subsp. *saccharum* (sugar maple) 110–11
    A. tataricum 314–15, 300
achenes 176, 300, 320, 321, 324
acidity levels 27, 155, 254
ackee (*Blighia sapida*) 335
Aconitum (monkshood) 189, *229*
actinomorphic flowers 188
aculeate leaves 133
acumen (drip tip) 134
Adenium obesum 189
aerenchyma 48
aerial roots 24, **38–39**, 84, 97
Aesculus hippocastanum (horse chestnut) 72, *293*
Aesculus pavia (red buckeye) 115
aggregate fruits *295*, 301
Aglonema sp. 151
Ailanthus altissima (tree of heaven) 136, 314
air plants **42–43**
Album of Fruit and Flowers, Chen Hongshou 194–95
Alcea sp. (hollyhock) **252–53**
alder (*Alnus*) *74*, 217
algae *12*
alkaline soils 27
allelopathic plants 221
Allium sp. 214–15, 216
Alnus sp. (alder) *74*, 217
Aloe sp. 143, 146
Aloidendron dichotomum (quiver tree) **146–47**
alternate leaf arrangements (phyllotaxis) *122*
alternate branches *70*
alternate buds 73

Amanita muscaria (fly agaric) **34–35**
Amazon water lily (*Victoria amazonica*) 166–67, 212–13
amber 82
amino acids 227
Amorphophallus titanum (titan arum) **210–11**
Ampelopsis glandulosa (porcelain berry) 122–23
anaerobic soil *48*, *53*
ancient gardens 296
ancient plants **186–87**
angel wings (*Caladium bicolor*) 151
angiosperms 15, 51, 186, 244, 286
Angraecum sesquipedale (Madagascan orchid) *229*
annual plants 191
ant fern (*Lecanopteris carnosa*) 88, 340–41
ant house plant (*Myrmecodia beccarii*) 89
"ant plants" **88–89**
anthers *184*, *185*, *218*
    after fertilization 191, 197
    in buzz pollination 234
    in grass plants *224*
    in pollination 207, 214, *232*, *233*, *256*, *259*
        cross-pollination *196*
        self-pollination 201, 245, *268*
Anthoceros sp. *14*
anthocyanins 138, 153, 154, *155*, 254
Anthurium sp. 38–39, 134–35, 217
Antirrhinum majus 189
ants **88–89**, 160, 327
aphids 266, 267
apical meristem 100
apple (*Malus domestica*) 301
aquatic bladderworts 170
aquatic plants 48, 51, 170, 333
    leaves **166–67**
    "primitive" *186*
    water lilies
        Nymphaea sp. **50–51**, 133
        Victoria amazonica (giant water lily) 166–67, 212–13
Aquilegia sp. 189, 228–29, 301
Arabidopsis thaliana (thale cress) 133
Araceae family (arum) 208–09
Araucaria araucana (monkey puzzle tree) 70–71
Arbutus sp. 294–95
Archidendron ramiflorum 76–77
Arctium sp. (burdock) *277*, 313
arils 289, 309
Aristolochia (Dutchman's pipe) 280
arrowhead (*Sagittaria sagittifolia*) 133
Artocarpus sp. 76, *181*
Arts and Crafts movement **116–17**

arum family (Araceae) 208–09
Asclepias syriaca (common milkweed) **322–23**
ash, common (*Fraxinus excelsior*) 73
Asian fig 135
Asparagaceae family (asparagus) 84
Asplenium australasicum 338
Asteraceae family (daisy) 218–19
auxin 122
awns 304, 319, *326*, 327
axillary buds 72

B

bacteria *33*, 67, 82, 135, 177
Balantium antarcticum (syn. *Dicksonia antarctica*) 90, 120
bamboo, moso (*Phyllostachys edulis*) 106–07
banana (*Musa* sp.) 90, **302–03**
banana plants *90*
Banks, Sir Joseph 92
Banksia sp. 242–43, 330–31
barberry (*Berberis* sp.) 79
bark 59, *64*, **66–67**, 68, 82
basal angiosperms 186–87
basal plate 86
bats 80, 100, **244–45**, 253, 302
Bauer, Franz 271
Bauhinia scandens 134
beach, common (*Fagus sylvatica*) 73
beehive ginger (*Zingiber spectabile*) 279
bees 199, 250
    plants for 158
    pollination by 199, 214, **232–33**, 264–65
        buzz pollination 235
        flower preferences of 231, 248, 249
        nectar guides for 250–51, 253
beetles *212*, 248
beetroot (*Beta vulgaris*) 22–23
Begonia sp. *134*, 150
Beltrami, Giovanni 117
Berberis sp. (barberry) 79
Berkheya purpurea (South African thistle) 160–61
berries 47, 123, 300, 302
    see also arils
Beta vulgaris (beetroot) 22–23
betalains 179
Betula sp.(birch) 59, 67, 333
biennial plants 22, **28–29**, 191
bifoliate leaves 115
bilabiate flowers 189
bilobed leaves 132

bird of paradise plant (*Strelitzia reginae*) 262–63, 309
birds
    in botanical art 195
    nesting sites for 54, 146
    pollination by 47, 100, 231, **240–41**, 248, 253, 262–63, 302
    seed preferences of 308, 309
    seeds spread by 305, **306–07**
Bismarckia nobilis (Bismarck palm) **164–65**
blackberry (*Rubus* sp.) 298–99
blades, leaf 113
Blechnum orientale 120
bleeding heart (*Lamprocapnos spectabilis*) 268–69
Blighia sapida (ackee) 335
blue flowers 27, 249
blue grama grass (*Bouteloua gracilis*) 304–05
blue Mauritius bellflower (*Nesocodon mauritianus*) 188, 226–27
Borago officinalis (borage) 234–35
Borassus flabellifer (toddy palm) 92
Boston ivy (*Parthenocissus* sp.) 97
botanical art
    pre 18th century **62–63,** 296–97
        herbals **140–41**
        in China **194–95**
    18th and 19th century **92–93, 270–71**
        Arts and Crafts movement **116–117**
        classification of plants **310–11**
        female artists **334–35**
        in Japan **204–05**
        in North America **280–81**
        Postimpressionism **30–31**
    20th century **124–25**
        modernist movement **236–37**
botanists 92, 141, 205, 311
bottlebrush (*Callistemon* sp.) 217, 243
Bougainvillea × *buttiana* 179
Bouteloua gracilis (blue grama grass) 304–05
bracken (*Pteridium* sp.) 114
bracts **176–77**, **178–79**, **278–79**
    hairy 28
    in air plants 43
    in grasses 222, 223, *224*
    in *Magnolia* 72, 74, 75, 186
    in ray and disc flowers 218
    of cones *283*, 288
    of seeds 313, 319
    scaly 231
    spiny 161, 277

branches
  arrangements **70–71**
  cauliflory **76–77**
  climbing plants **94–95**, **96–97**
  defences **78–79**
  in winter **72–73**, **74–75**
  see also stems
breadfruit (*Artocarpus altilis*) 76
Brillantaisia lamium 189
bristles 100
Brittany, McEwen 125
broadleaf trees 71, 110–11, 118
Bromeliaceae family (pineapple) 42–43, 168–69
broom (*Cytisus scoparius*) 327
bryophytes 14
bud scales 72–73, 75, 119, *181*
buds 72–73
  defences **274–75**
  flowers 247
  insulated **74–75**
  leaves 118
  see also flower buds
bulbils 336, 337
bulbs 36–37, **86–87**, 337
bumblebees *see* bees
burdock (*Arctium* sp.) *277*, 313
burrs 277, 313
butcher's-broom (*Ruscus aculeatus*) 60–61
buttercup (*Ranunculus acris*) 217, 328
buttercup family (Ranunculaceae) 328
butterflies 248, 249, 253, 322
butterfly vine (*Mascagnia macroptera*) 265
buttress root systems *24*
buzz pollination **234–35**

C

cacti 79, **98–99**, 103, 180, 199
  Mammillaria infernillensis 98–99, 160
  queen of the night (*Epiphyllum oxypetalum*) **246–47**
  saguaro cactus (*Carnegiea gigantea*) 98–99, **100–01**
Caesalpinia sappan (Indian redwood) 92
calabash (*Crescentia cujete*) 76, *244*
Caladium bicolor (angel wings) 151
Calathea bella 150
calcarate flowers 189
calceolate flowers 189
Callistemon sp. (bottlebrush) 217, 243
Caltha palustris (marigold) 250
Calystegia sepium (hedge bindweed) 94–95
calyx 176, *218*, 293, *321*
CAM (crassulacean acid metabolism) photosynthesis *142*
cambium *64*, 67
campanulate flowers 188
Campbell's magnolia (*Magnolia campbellii*) 72

Cannabis sp. (cannabis) 115
Cape sundew (*Drosera capensis*) **172–73**
Cape weaver bird 263
capitulum inflorescences 217
capsules 301, 312–13, **324–25**, 326, 328
  honesty plants 287
  spores 343
  see also follicles
carbohydrates 22, *29*, 34, 44, 60, 112
  see also photosynthesis
carbon dioxide 54, 102, 112, *113*, 126, 128, *129*
  in succulent plants 142
  see also photosynthesis
cardoon (*Cynara cardunculus*) 176–77
Carica papaya (papaya) 76, 334
Carnegiea gigantea (saguaro cactus) 98–99, **100–01**
carnivorous plants **170–71**, **172–73**, **256–57**
carotenoids 154, *155*
carrion flies 261
carrion flower (*Amorphophallus titanum*) 210–11
carrot, wild (*Daucus carota*) 28–29, 216
Carya ovata (hickory) 66, 115
caryopsis fruits 301
Cassia fistula (golden rain tree) 216
Castanea sativa 66
castor oil plant (*Ricinus communis*) 309
caterpillars 157, 161
catkins 217, 222
Caucasian oak (*Quercus macranthera*) 137
cauliflorous plants **76–77**
Cedrus atlantica (cedar) 282–83, 288
Ceiba sp. 66
  C. pentandra (kapok tree) **80–81**
Cenchrus sp. 224, 225
Cercidiphyllum japonicum (katsura) 136
Cercis siliquastrum (Judas tree) 76
Chasmanthium latifolium (northern sea oat) 222–23
chasmogamous flowers 268
Chen Chun 194
Chen Hongshou 194–95
cherry trees *see Prunus* sp.
chimeras 150
Chinese botanical art **194–95**
Chinese lantern (*Physalis* sp.) 292–93
Chinese pink (*Dianthus chinensis*) 217
chlorophyll 27, *119*, 128, 154, *155*
Chlorophytum comosum (spider plant) 104
chloroplasts *119*, *129*, *142*, 154
Chrysolepis chrysophylla (golden chinkapin) 132
Cibotium glaucum 121
Cichorium intybus 198
ciliate leaves 136
circinate vernation 120
Cirsium rivulare (plume thistle) 231
Citrus sp. 79, 115, 300
class (classification) 16
classification of plants **16–17**, 311

cleft leaves 137
cleistogamous flowers **268–69**
Clematis sp. 60–61, 188, 320–21
climate *see* weather extremes
climbing plants 58, **94–95**, **96–97**, *181*
clones 36, *174*, **336–37**
  aspen groves 68
  bamboo 106
  banana plants 302
  dandelion 317
  dragon tree 85
clover (*Trifolium* sp.) 32–33, 115
clustered buds 73
coastal habitats 24, 53, 54
coats (seeds) 288, *293*, 309, 333
cocoa plants (*Theobroma cacao*) 76
coconut *333*
coconut palm (*Cocos nucifera*) 162–63
Cocos nucifera (coconut palm) 162–63
Codiaeum variegatum 150
coir *333*
Colocasia esculenta 112, 131
colonies *see* clones
colour pigments 154–155, **248–249**, *254*
  for pollination **254–55**, **278–79**
  green (chlorophyll) *118*, 154
    see also chlorophyll
  red 151, 153, 154
    see also anthocyanins
  yellow and orange (carotenoids) 154, 155 248
colpus 198
common ash (*Fraxinus excelsior*) 73
common beach (*Fagus sylvatica*) 73
common foxglove (*Digitalis purpurea*) 232–33
common hollyhock (*Alcea rosea*) **252–53**
common ivy (*Hedera helix*) 58, 97
common lime (*Tilia × europaea*) 73
common milkweed (*Asclepias syriaca*) **322–23**
common nettle (*Urtica dioica*) 156–57
common teasel (*Dipsacus fullonum*) 276–77
compound corymb inflorescences 216
compound cyme inflorescences 217
compound leaves **114–15**
compound umbel inflorescences 216
cones 71, **282–83**, 288, 289, **290–91**
conifers
  branch arrangement 70, 71
  leaves 118
  seeds and cones 283, 288, 289, *291*, 331
contractile roots **36–37**
cordate leaves 132
corkscrew hazel (*Corylus avellana* 'Contorta') 59
corkscrew plants 170
corky bark 66
corms 36, *87*, *104*, 211
corn (*Zea mays*) 301

corollas 184, **188–89**, *212*, 240, *268*, 295
coronate flowers 188
corpse flower (*Rafflesia arnoldii*) **260–61**
corpse plant (*Amorphophallus titanum*) 210–11
cortex 20, 33, 36, *39*
Corylus sp. (hazel) 59, 222, 301
Costus guanaiensis 126–27
Cotinus coggygria (European smoketree) 132
cotton (*Gossypium*) 321
cotyledons *286*, *319*, *324*
cranesbill (*Geranium sanguineum*) 326–27
Crassula rupestris subsp. *Marnieriana* (jade necklace plant) 142–43
crassulacean acid metabolism (CAM) *142*
Crataegus sp. (hawthorn) 79
crenate leaves 136
crenulate leaves 136
Crescentia cujete (calabash) 76, *244*
Crocosmia 104
crop-rotation systems 32
crops 22, 23, 29, 32, 244
cross-pollination *196*, **200–01**, **208–09**, 214, 257
  of cones 282
  scent for 258
crown of thorns plant (*Euphorbia milii*) 180–181
cruciform flowers 188
Cucumis metuliferus 300
culm 106, 223
Culpeper, Nicholas 141
cultivar (classification) 16
cupule bracts 178
cuticles 61, 112, 134, 142, 145
  in succulent plants 142
Cyathea sp. 15
cycads 120, 283
cyme inflorescences 217
Cynara cardunculus (cardoon) 176–77
Cypress trees, Kanō Eitoku 31
cypsela fruits 300, *321*
Cytisus scoparius (broom) 327

D

da Vinci, Leonardo 63
Dactylorhiza praetermissa 198
daffodil (*Narcissus × odorus*) 271
Dahlia 'David Howard' 192
daisy family (Asteraceae) 218–19
dandelion (*Taraxacum* sp.) 23, 217, 218, **316–17**, 300, *321*
Darlingtonia (pitcher plant) *257*
Darwin, Charles 172, *229*
Datura wrightii (jimson weed) 237
Daucus carota (wild carrot) 28–29, 216
De Materia Medica, Dioscorides 140, 141

defences 138
 bracts for 176
 of buds **274–75**
 of flowers **276–77**
 of leaves **160–61**
 of stems **78–79**, 80–81,
 poisons and irritants 221, 293
 sap 103, 180, 231, 322
 prickles and thorns **78–79**, 80–81,
 158, 166, 298
 see also hairs; spines
deforestation 211, 261
dehiscent seeds 324
Delphinium sp. 216
deltoid leaves 133
dentate leaves 137
denticulate leaves 137
desert habitat 100, 102, 158, 244
 see also cacti; succulent plants
determinate inflorescences 216–17
devil's claw plant (*Ibicella lutea*) 312–13
devil's tobacco plant (*Lobelia tupa*)
 240–41
Dianella tasmanica (Tasman flax-lily)
 306–07
Dianthus chinensis (Chinese pink) 217
Dichorisandra sp. *235*
Dicksonia antarctica (syn. *Balantium
 antarcticum*) tree fern *90, 120*
Dieffenbachia sp. (dumb canes) 208
digestion (carnivorous plants) *171*,
 172
Digitalis purpurea (common foxglove)
 232–33
dioecious plants 165, 207
Dioscorea sansibarensis 134
Dioscorides, Pedanius 140, 141
Diphasiastrum digitatum 14
Diplazium proliferum 336–37
diploid cells 291
Dipsacus fullonum (common teasel)
 276–77
disc flowers **218–19**, 220
Dischidia 88
division (classification) 16
Dodecatheon sp. *235*
dog rose (*Rosa canina*) 188, 193, 270,
 274–75
dog violet (*Viola riviniana*) 269
domatia *88*
double samaras *see* samaras
double serrate leaves 136
Douglas fir (*Pseudotsuga menziesii*) 111,
 288
Dracaena draco (dragon tree) **84–85**
drip tips **134–35**
Drosera sp. (sundew) 170, **172–73**
droughts 37, 64, 98, 138, 146, 158
drupe fruits 299, 301
dry fruits 300–01
Dryopteris filix-mas 120
Duchesnea indica, syn. *Fragaria indica*
 (mock strawberry) 105

dumb canes (*Dieffenbachia* sp.) 208
Dürer, Albrecht 62, 63
Dutch East India Company 205
 see also East India Company
Dutchman's pipe (*Aristolochia*) 280

E

early dog-violet (*Viola reichenbachiana*)
 136
East India Company 92, 311
 see also Dutch East India Company
Echinacea sp. 217, 218–19
Echinops bannaticus 'Taplow Blue' (globe
 thistle) 192
Egyptian gardens 296
Ehret, Georg Dionysius 310, 311
Eitoku, Kanō 31
elaiophores 265
elaiosomes 327
elliptic leaves 133
Embothrium coccineum (Chilean fire bush)
 132
endocarp 294, *333*
endosperm *286, 333*
English Physitian, The, Culpeper 141
entire leaves 136
epicalyx bracts 178
epicarp 294
epicotyls *286*
epidermis 39, 112, *275*
Epiphyllum oxypetalum (queen of the
 night) **246–47**
epiphytes 67, 80, 88, 247
 air plants 42–43
 pearl laceleaf 38–39
 strangler fig 41
Equisetum (horsetails) 48–49
Erodium cicutarium (stork's bill) 300,
 326
Eruca sativa 188
Espostoa fruitescens 244
Eucalyptus sp. 66, 148–49, 243
Eucomis pole-evansii 178
Eucryphia sp. 131, 132, 136
eudicots 15, 186
 leaves *181*
 roots 22
 stems **60–61**, 67
Euphorbia sp. 102–03, 145, 178, 198
 E. milii (crown of thorns plant)
 180–181
European pear tree (*Pyrus communis*)
 125
evaporation
 in aquatic plants 167
 in succulent plants 99, 142
 prevention of 112, *126*, 138, 148,
 180
evergreen plants 110, 111, 190
everlasting pea blooms (*Lathyrus
 latifolius*) 202–03

evolution, of plants 133, 186–87
excrement, seed dispersal by 305,
 306–07
exine 198
exocarp *333*
exodermis *39*
eyelash grass (*Bouteloua gracilis*)
 304–05

F

Fagus sylvatica (common beach) 73
false pepperwort (*Marsilea crenata*) 115
family (classification) 16
fan palm (*Livistona mauritiana*) 93
feathertop (*Cenchrus longisetus*) 224–25
female parts, of flowers 51, 184, 212
 pistils 184, *196*, 197, *259*, *299*
 style *184*, 185, *196*, *256*, *299*, 328
 see also ovaries; ovules; stigma
fenestrate leaves 133
fennel (*Foeniculum vulgare*) 328
fennel flower (*Nigella sativa*) **328–29**
ferns 14, 15, *90*, 114, **120–21**, 337
 spores **338–39**, **340–41**
fertilization **196–97**, 207, 214, 268
 ferns *340*
 gametophytes *340, 343*
 grasses 222, 224
 gymnosperms 282–83, 290, *291*
 pin and thrum flowers *200*
 see also pollination
fertilizers 27
fibrous roots **20–21**, 22
fibrous tree trunks **90–91**
Ficus sp. (fig) 76, 133, 135, *263*
 strangler fig **40–41**
fiddleheads 120, 337
fig (*Ficus* sp.) 76, 133, 135, *263*
filaments 74, *184*, 185, *207*, *235*, 317,
 322, 341
fire, release of seeds by 68, **330–31**
 see also wildfires
firethorn (*Pyracantha*) 79
fish 54, 166
fissured bark 67
Fittonia sp. 150
flaking bark 66
flies 211, *229*, 231, 248, *259*, 272
 carrion flies 261
 hoverflies 266
floating leaves **166–67**
Flora Indica, Roxburgh 92
florets 192, 214, 218, 231
 grasses 222–223, *224*
florigen *193*
flower buds 72, **74–75**, 77, 86–87, 247
 corpse flower 261
 defences **274–75**, **276–77**
 "primitive" 186
 sunflowers 221
 see also buds

evolution, of plants 133, 186–87

flowers 13, **184–85**, 218
 and fruit formation 298, 299
 carnivorous plants **256–57**
 closure at night **272–73**
 colour pigments **248–49**, 250–51,
 253, **254–55**
 defences **276–77**
 development of **190–91**, **192–93**
 fertilization of **196–97**, *199*,
 206–07
 flowerhead *see* inflorescences
 hermaphroditic **212–13**
 in grass plants **222–23**, **224–25**
 in succulent plants 143, 247
 nectar **226–27**, **228–29**, 250–51,
 253
 pollination **202–03**, **266–67**,
 268–69
 by mammals **242–243**,
 244–45,
 cross-pollination **208–09**,
 see also bees, pollination by;
 birds, pollination by; bracts
 "primitive" species 186–87
 scent 211, **258–59**, 261
 shapes of **188–89**, 203, 221, 231
 winter flowering plants **238–39**
 see also cones; flower buds;
 inflorescences
Flowers, Warhol 236
fly agaric (*Amanita muscaria*) 34–35
flying foxes 244
 see also bats
Foeniculum vulgare (fennel herb) 328
foliaceous stipules *181*
follicles 301, 308, 324, 331
 milkweed 322
 nigella 328
food *see* nutrients
Foord, Jeannie 117
forced pollination 258
forests habitats 41, 76, 158, 331
 bamboo, moso (*Phyllostachys edulis*)
 106–07
 quaking aspen (*Populus tremuloides*)
 68–69
 see also mangroves; rainforest
 habitats
Forsythia viridissima 74
fossils 82, 137
foxglove, common (*Digitalis purpurea*)
 232–33
Fragaria × ananassa (edible strawberry)
 104, 300
Fraxinus excelsior (common ash) 73
fronds 114, 134, 162, 163, 337,
 338
 palm trees 165
 unfurling (fiddleheads) **120–21**
frost damage 131
fructose 227
fruit bats 244
fruitlets 299

fruits 165, 286, **294–95**, **298–99**, **300–01**
  figs 41
  witch hazel 239
  see also seeds
Fruits and Flowers of Java, The, van Nooten 334
fuchsia sp. 184–85
fungi *12*, 34, 67, 82, 135, 177

G

Galanthus elwesii (giant snowdrop) *238*
galeate flowers 189
gametes 291, *291*, 340
gametophyte *340*, 341, 343
gardens **296–97**
garlic, wild (*Allium ursinum*) 216
geckos 227
genus (classification) 16
Geranium sanguineum (cranesbill) 326–27
germination 51, 305, 306, *324*, 333, 341
giant hogweed (*Heracleum mantegazzianum*) 230–31
giant redwood (*Sequoiadendron giganteum*) 111, 288
giant rhubarb (*Gunnera manicata*) 131
giant snowdrop (*Galanthus elwesii*) *238*
giant water lily (*Victoria amazonica*) 166–67, 212–13
Ginkgo biloba (maidenhair tree) 132, 283, 289
globe artichoke 176
globe thistle (*Echinops bannaticus* 'Taplow Blue') 192
glochids 98
glucose *129*, 227
glume bracts 178
gnats 259
Gogh, Vincent van 30–31
golden columbine (*Aquilegia chrysantha*) 228–29
golden rain tree (*Cassia fistula*) 216
Gossypium (cotton) 321
grapple plant (*Harpagophytum procumbens*) 313
Graptoveria 174
grasses 22, 58, 62, 106
  flowers **224–25**
  pollination by wind **222–23**
  seed dispersal 304–05
Great Piece of Turf, Dürer 62, 63
green hellebore (*Helleborus viridis*) *249*
green pigment (chlorophyll) *118*, 154
  see also chlorophyll
greenhood orchids (*Pterostylis* spp.) 258–59
Grevillea 'Coastal Sunset' 241
growth rings **64–65**
guard hairs 232
Gunnera manicata (giant rhubarb) 131

Guzmania 'Olympic Torch' 178
gymnosperms 14, 15, **282–83**, 286, 288–89, 290–91
gynostemium *259*
Gynura aurantiaca (velvet plant) 138

H

haircap moss (*Polytrichum* sp.) 14, **342–43**
hairs
  bracts 176
  foliage 96, 99, **138–39**, 176, 233
  seeds **320–21**
  silky hairs 43, 75, 138, **320–21**
  see also trichomes
Hamamelis sp. (witch hazel) 74, 238–39
Hanging Gardens of Babylon 296
hanging lobster claw plant (*Heliconia rostrata*) 278–79
haploid cells 291
hare's tail grass (*Lagurus ovatus*) 225
Harpagophytum procumbens (grapple plant) 313
hastate leaves 133
haustoria 44, 47
hawthorn (*Crataegus* sp.) 79
heart fern (*Hemionitis arifolia*) 339
heartwood *64*
Hedera sp. (ivy) 132, 133, *229*
  H. helix (common ivy) 58, 97
hedge bindweed (*Calystegia sepium*) 94–95
Heliamphoa (pitcher plant) *257*
Helianthus sp. (sunflower) **220–21**
Heliconia rostrata (hanging lobster claw plant) 278–79
helicopter samaras 153, 300, **314–15**
Helleborus sp. 190, 228, *229*, 249
Hemionitis arifolia (heart fern) 339
hemiparasitic plants 44
Heracleum mantegazzianum (giant hogweed) 230–31
herbaceous perennial plants 190
herbals **140–41**
herbicides 138, 322
herbivores
  defences against 48, 85, 153, 322
    hairs 120, 138, 156
    spines and prickles 70, 78, 100, 103, 161, 176, 180–81
  plants as food for 111, 158, 305
  seed preferences of 309
  seeds spread by 305, 312–13
Heritiera littoralis (looking-glass mangrove) 24–25
hermaphroditic flowers **212–13**, 278
hesperidium fruits 300
Hibiscus sp. 178, 200–01
hickory (*Carya ovata*) 66, 115
hip fruits 275, 300
hogweed, giant (*Heracleum mantegazzianum*) 230–31

Hoheria glabrata (ribbonwood) 133
Hokusai, Katsushika 205
holly (*Ilex* sp.) 111, 133, 137, 206–07
hollyhock (*Alcea* sp.) **252–53**
holoparasitic plants 44
honesty plant (*Lunaria annua*) 286–87
honeyeaters 241
honeysuckle (*Lonicera periclymenum*) 254–55
Honzo Zufu, Iwasaki Tsunemasa 204
hornworts 14
horse chestnut (*Aesculus hippocastanum*) 72, *293*
horsetails (*Equisetum*) 48–49
Hortus Cliffortianus, Linnaeus and Ehret 311
hosts *see* epiphytes
hummingbirds 241, 278
Humulus lupulus 178
Hungarian oak (*Quercus frainetto*) 72
Hyacinth 36–37, 86–87
Hydnophytum 88
Hydrangea sp. 26, 27, 216
hybridize *see* cross-pollination
hydrochory 333
hygroscopic movement *326*
Hymenaea courbaril (Indian locust tree) 115
Hypericum pseudohenryi (St. John's wort) 196–97
hyphae 34
hypococtyls *286*
hypocrateriform flowers 189
hysteranthy *74*

I

Ibicella lutea (devil's claw plant) 312–13
Icelandic poppy (*Papaver nudicaule*) 58, *231*
Ilex sp. (holly) 111, 133, 137, 206–07
Illicium anisatum (star anise) *186*
illustrations *see* botanical art
imperfect flower structures *207*
incised leaves 137
indehiscent seeds *293*, 324
indeterminate inflorescences 216–17
India, flora in 92
Indian locust tree (*Hymenaea courbaril*) 115
Indian Onion painted in Benares, McEwen 124
Indian redwood (*Caesalpinia sappan*) 92
inflorescences 177, 211, **214–15**, **216–17**
  see also bracts; flowers
infundibuliform flowers 189
insects
  "ant plants" **88–89**
  beetles *212*, 248
  butterflies 248, 249, 253, 322
  carnivorous plants **170–71**, **172–73**, **256–57**

insects *continued*
  wasps *228*, 248, 249
    fig wasps 41, *263*
  see also bees; flies
insulation, buds **74–75**
internodes 58, 96
involucral bracts 178
Ipomoea purpurea (morning glory) 132, 189
Ipomopsis aggregata 189
Iris sp. 217
iron deficiency 27
ivy (*Hedera* sp.) 132, 133, *229*
  common ivy (*H. helix*) 58, 97
Iwasaki Tsunemasa 204

J

Jacinthe Double, Spaendonck 271
jade necklace plant (*Crassula rupestris* subsp. *Marnieriana*) 142–43
Japanese botanical art 30, 31, **204–05**
Japanese lantern (*Hibiscus schizopetalus*) 200–01
Japanese maple (*Acer palmatum*) 136, **152–53**
jimson weed (*Datura wrightii*) 237
Jimson weed, O'Keeffe 237
Judas tree (*Cercis siliquastrum*) 76
Juglans sp. (walnut) 72, 132
Juniperus sp. (juniper) 289

K

kaffir lime (*Citrus hystrix*) 115
Kalanchoe daigremontiana (mother of thousands) 174–75
Kan-en 205
Kanō Eitoku 31
kapok tree (*Ceiba pentandra*) **80–81**
katsura (*Cercidiphyllum japonicum*) 136
Koelreuteria paniculata (pride of India) 137
König, Johan 92

L

labellum 258
Laelia tenebrosa, Philodendron hybrid…, Sellars 124
Lagurus ovatus (hare's tail grass) 225
Lambdin, George Cochran 281
lamb's ears (*Stachys byzantina*) 138–39
lamina 118
Lamprocapnos spectabilis (bleeding heart) 268–69
lanceolate leaves 132
"Large Flowers", Hokusai 205
Larix sp. 15
Lathraea clandestina (purple toothwort) 45

Lathyrus sp. 189, 202–03, 301
leaf arrangement 123
leafless stems **102–03**
leaflets 341
leafy bracts 178
leaves **110–11**, **112–13**, 129
    and photosynthesis **128–29**
    and plantlets **174–75**
    and transpiration **126–27**
    aquatic plants **166–67**, *167*
    arrangement of **122–23**, 186
    carnivorous plants **170–71**, 172
    colour pigments 153, **154–55**
    compound **114–15**
    defences **138–39**, **156–57**,
        160–61, **180–81**
        see also bracts; trichomes
    development of **118–19**, **120–21**,
        245
    in arts and crafts design **116–17**
    in weather extremes **162–63**, 165
    on succulent plants **142–43**
    shapes and margins **132–33**,
        134–35, **136–37**
    size **130–31**
    variegated **150–51**
    waxy surfaces of **144–45**, **148–49**,
        165
Lecanopteris carnosa (ant fern) *88*, 340–41
legumes 33, 301
lemma bracts 178, 222, *224*
lemon (*Citrus* × *limon*) 300
lemon balm (*Melissa officinalis*) 133
lemurs 242
lenticellate lark 66
lenticels *53*, 67
Leucaena leucocephala (white leadtree) 115
Leucojum vernum (spring snowflakes) *254*
lichens *12*
ligules *218*
lilac (*Syringa* sp.) 216
Lilium sp. 15, 198, 324–25
lime, common (*Tilia* × *europaea*) 73
lime, kaffir (*Citrus hystrix*) 115
linear leaves 132
Linnaeus, Carl 16, 17, 92, 172, 311
Liquidambar styraciflua (sweet gum) 132
Liriodendron tulipifera 67
liverworts 14
Livistona mauritiana (fan palm) 93
lobate leaves 137
Lobelia tupa (devil's tobacco plant)
    240–41
London plane (*Platanus* × *hispanica*) 73
Lonicera periclymenum (honeysuckle)
    254–55
loofah (*Luffa cylindrica*) 96–97
looking-glass mangrove (*Heritiera*
    littoralis) 24–25
lotus (*Nelumbo* sp.) 17, 144, 272–73,
    332–33
love-in-a-mist (*Nigella damascena*) 328
Luffa cylindrica (loofah) 96–97

Lunaria sp. 286–87, 301
lungwort (*Pulmonaria officinalis*) 254
Lycianthes rantonnetii 188
lycophytes 14
Lysichiton americanus 178

M

Maclura pomifera 301
Madagascan orchid (*Angraecum*
    sesquipedale) 229
madrone (*Arbutus menziesii*) 294
Magnolia 74–75
    M. campbellii (Campbell's magnolia)
        72
    M. grandiflora 15, 186–87
    M. macrophylla (bigleaf magnolia)
        132
    M. × *soulangeana* 'Rustica Rubra'
        (saucer magnolia) 308
Magnoliid family (magnolia) 15,
    186–87
maize (*Zea mays*) 301
male parts, of flowers 51, 184, 212
    filaments 74, *184*, *185*, *207*, *235*,
        317, 322, 341
    see also anthers; stamens
mallow family (Malvaceae) 253
Malus domestica 301
Malvaceae family (mallow) 253
mammals
    defences against 138, 157, 161, 176,
        322
    pollination by **242–43**
    seeds spread by 307, 309, 312–13
Mammillaria infernillensis (cactus) 98–99,
    160
Manchurian walnut (*Juglans mandshurica*
    var. *sieboldiana*) 72
mangroves 24, **52–53**, 244
    Rhizophora sp. (mangrove) **54–55**
maples 90, 110, 136, 152–53, 314
Marcgravia evenia *244*
Marchantia polymorpha 14
margins, leaf **136–37**
marigold (*Caltha palustris*) 250
marsh woundwort (*Stachys palustris*) 217
Marsilea crenata (false pepperwort) 115
Mascagnia macroptera (butterfly vine) 265
McEwen, Rory 124–25
medicinal plants 85, 120, 328, 339
    herbals 141
Mee, Margaret 334
Melica nutans 178
Melissa officinalis (lemon balm) 133
Mentha sp. (mint) 136
mesocarp 294
mesophyll cells 112, *113*, *129*, *142*
micropyle *291*
midrib vein 113
milkweed, common (*Asclepias syriaca*)
    322–23

mimicry
    as defence 151
    for pollination 176, 211, 259, 261
        orchids 266–67, **264–65**
minerals 20, *21*, 27, 60
mint (*Mentha* sp.) 136
mistletoe (*Viscum album*) 44, **46–47**
mock strawberry (*Duchesnea indica*, syn.
    Fragaria indica) 105
modern art **236–37**
monkey puzzle tree (*Araucaria araucana*)
    70–71
monkshood (*Aconitum*) 189, *229*
monocot plants 15, 85
    flowers 184
    leaves *113*
    seeds *286*, *324*
    stems 60–61, 67
monoecious plants *207*
Monstera deliciosa (Swiss cheese plant)
    133, *134*
mophead hydrangea (*Hydrangea*
    macrophylla) 27
Morris, William 116–17
Morus rubra (red mulberry) 137
moso bamboo (*Phyllostachys edulis*)
    106–07
mosses (*Polytrichum* sp.) 14, **342–43**
mother of thousands (*Kalanchoe*
    daigremontiana) 174–75
moths *229*, 238, 247, 248
Mount Fuji with Cherry Trees in Bloom,
    Hokusai 205
multifid fan leaves 133
multifoliate leaves 115
multiple fruit *295*, 301
Musa sp. (banana plant) *90*, **302–03**
mushrooms (fungi) 34
mutualism 88, **262–63**
    see also symbiotic relationships
mycelium 34
mycorrhizae 34
Myristica fragrans (nutmeg) 334
Myrmecodia beccarii (ant house plant) 89
Myrmecophila 88

N

naked seeds **288–89**
Narcissus sp. 188, 271
nasturtium (*Tropaeolum majus*) 133
nectar 214, **226–27**, **228–29**, 255
    animals feeding on 242–43, 244,
        256, 323
    in cacti 100
    in carnivorous plants 171
    in quiver trees 146
    in sundews 172
    see also nectar guides; nectaries
nectar guides 253, **250–51**
see also colour pigments
nectaries *227*, 228–29, 241, 242

needles (leaves) 111
Nelumbo sp. (lotus) 17, 144, 272–73,
    332–33
Neoregelia cruenta 168–69
Nepenthaceae 170–71
Nesocodon mauritianus (blue Mauritius
    bellflower) 188, 226–27
nettle, common (*Urtica dioica*) 156–57
Nigella sp. 190–91, 328
    N. sativa (nigella) **328–29**
nitrogen *27*, **32–33**, 170, 172
nitrogenase 33
nitrogen-fixing plants **32–33**
non-flowering plants 14
Nooten, Berthe Hoola van 334
North, Marianne 334, 335
North American botanical art 280–81
northern sea oat (*Chasmanthium*
    latifolium) 222–23
nutmeg (*Myristica fragrans*) 334
Nutmeg foliage, flowers and fruit, North 334
nutrients
    absorption of *21*, *23*, **26–27**, 34
        in epiphytes 41, 47
        in leaves 112, 113, 126, 175
        in aquatic plants 166
        in carnivorous plants 170, *176*, *257*
    storage of **86–87**, 158, 286
nuts *293*, 301
Nymphaea sp. (water lily) 15, 16, **50–51**,
    133

O

oak (*Quercus* sp.) 82–83, 133, 178
    caucasian oak (*Q. macranthera*) 137
    cork oak (*Q. suber*) 66
    Hungarian oak (*Q. frainetto*) 72
    pin oak (*Q. palustris*) 137
    sessile oak (*Q. petraea*) 137
    strandzha oak (*Q. hartwissiana*) 307
    Turkey oak (*Q. cerris*) 137
obcordate leaves 132
oblanceolate leaves 132
oblong leaves 132
obovate leaves 132
Odontites vulgaris (red bartsia) 44
O'Keeffe, Georgia 236–37
Oncidium 264–65
operculum 171
opposite leaf arrangements *122*
opposite buds 73
Opuntia phaeacantha (prickly pear) 98
orange pigments (carotenoids) 154,
    155, 248
orange tree (*Citrus*) 79
orchids 43, 88, 265
    greenhood orchids (*Pterostylis* spp.)
        258–59
    Madagascan orchid (*Angraecum*
        sesquipedale) 229
    Oncidium 264–65

orchids *continued*
    Rothschild's slipper orchid
        (*Paphiopedilum rothschildianum*)
        266–67
    Vanda 13
order (classification) 16
Oriental fountain grass (*Cenchrus orientalis*) 225
ovaries *184*, 185, *207*, *218*, *224*, 268
    after pollination 274, 294
    fertilization of *196*, 197, *200*
    in carnivorous plants 257
    nectaries in *227*, *233*, 240
    seeds and fruits from 286, 294, *295*, *299*
    see also ovules
ovate leaves 133
ovules *184*, 185, *186*, 197, 289, *295*
    of cones *283*, *291*
ox tree (*Olyscias maraisiana*) 227
oxygen 20, 48, 128, *129*
    see also photosynthesis

P

Paeonia peregrina (peony) 190
paleas 178, *224*
palmate compound leaves 132
palmate leaves 114, 153
palmately lobed leaves 132
palms (*Cocos nucifera*) 162–63
Pando aspen grove, Utah 68, *337*
    quaking aspen (*Populus tremuloides*) 68–69
panicle inflorescences 216, 225
pansy (*Viola*) 327
Papaver sp. (poppy) 58, *231*, 301
papaya (*Carica papaya*) 76, 334
papery bark 67
papery bracts 179
Paphiopedilum (slipper orchid) 266–67
papilionaceous flowers 189
pappus 176, *218*
parachutes (seeds) **318–19**, *321*
parasitic plants 41, **44–45**, 47, 261
parenchyma cells 143
Parthenocissus sp. *97*
"paw-dripping" 242
peach (*Prunus persica*) 301
pearl laceleaf (*Anthurium scandens*) 38–39
pedicels *184*, 185, 200, 216, 314, 332
peduncles 58, 214, 216, 268
peeling bark 66
peltate leaves 133
peony (*Paeonia peregrina*) 190
pepo fruits 300
perennial plants *190*, 191, *238*
pericarp 294, 314
Persian gardens 296
personate flowers 189
petaloid bracts 178

petals *184*, 185, 272–73, 328
    closed flowers 268
    for pollination 196, 203, *212*, 214, 226, 259, 264
    in flower development 74, 190, 191, 197
    ray flowers 218, 220, 221
    winter flowers 238
    see also bracts
petiole 79, 134, 145, 160, *181*, 339
petunia, wild (*Ruellia chartacea*) 279
Phalaenopsis sp. 188
Philodendron ornatum (philodendron) 128–29, 137
Phleum pratense 198
phloem *39*, 60, *64*, 112
phosphorus *27*, 170
photoperiodism 193
photosynthesis 112, 128, *129*, 136
    by aerial roots 38
    chlorophyll in *118*
    crassulacean acid metabolism (CAM) 142
    in bulbous plants 86
    in ferns 120
    in rainforest habitats 135
    in variegated leaves 150
    leaf size for 131
    of succulent plants 102
Phragmipedium sp. 189
Phyllostachys edulis (moso bamboo) **106–07**
phyllotaxis 123
Physalis sp. (Chinese lantern) 292–93
Phytanthoza iconographia, Weinmann 310
phytotelmata 168
pin flowers *200*
pin oak (*Quercus palustris*) 137
Pineapple (*Ananas sativus*), Ehret 311
pineapple family (Bromeliaceae) 42–43, 168–69
pink pigment (flowers) 27, 248, 254
pinnae 114, *134*, 341
pinnate compound leaves 132
pinnate leaves 114, 134, 162
pinnatifid leaves 133
Pinus sp. 66, 83, 131, 198, 288
pistils *184*, *196*, 197, *259*, *299*
Pisum sativum 181
pitcher plants 170, *171*, **256–57**
pith cells *39*
plane tree (*Platanus* sp.) 16, 66, 73
plant hunters 334
plantlets **174–75**, 336, 337
Plants of the Coast of Coromandel, Roxburgh 92
Platanus sp. (plane tree) 16, 66, 73
plume thistle (*Cirsium rivulare*) 231
pneumatophores *53*
pods **324–25**
poinsettias 278
poisonous plants 34, 47, 102, 138, 156, 221, 293, 309, 322

pollen cones 71, 282, 289
pollen grains 196, **198–99**, **202–03**
pollination
    attraction for pollinators 76, 211, **230–31**, 253
        colour **248–49**, 278–79
        see also nectar
    in carnivorous plants 172, *256*
    methods
        by animals 41, *212*, 229, **242–43**, 247, *263*
            buzz pollination **234–35**
            restricted access 234, *235*
            see also bats; bees, pollination by; birds, pollination by
        by deception **266–67**
        by wind 71, 153, 317, 319, 321
            cones 291
            grasses **222–23**, 224, 225
    see also cones; cross-pollination; fertilization; self-pollination
pollinia 322
Polypodium sp. 339
Polyscias maraisiana (ox tree) 227
Polytrichum sp. (haircap moss) 14, **342–43**
pome fruits 301
poppies (*Papaver nudicaule*) 58, *231*
Populus tremuloides (quaking aspen) **68–69**
porcelain berry (*Ampelopsis glandulosa*) 122–23
porose capsule fruits 301
possums 243
Postimpressionist art 30–31
potassium *27*
powder coverings 146
prickles 78–79, 80–81, 166
prickly pear (*Opuntia phaeacantha*) 98
pride of India (*Koelreuteria paniculata*) 137
"primitive" species **186–87**
Primrose *200*
Primula vulgaris 188
prop root systems *24*, 54, 55
protandry *212*
proteins 33
protogyny *212*
Prunus sp. 66, 73, 136, 301
Pseudobombax ellipticum (shaving brush tree) 244–45
Pseudodracontium lacourii 209
pseudofruits 300
pseudo-terminal buds 72
Pseudotsuga menziesii (Douglas fir) 111, 288
Ptelea trifoliata (hoptree) 132
Pteridium sp. (bracken) 114
Pterostylis spp. (greenhood orchids) 258–59
Pucibotium glaucum 121
Pulmonaria officinalis (lungwort) 254
purple coneflower (*Echinacea purpurea*) 217
purple pigments 155, 249

purple toothwort (*Lathraea clandestina*) 45
Pyracantha (firethorn) 79
Pyrus communis (European pear tree) 125

Q

quadrifoliate leaves 115
quaking aspen (*Populus tremuloides*) 68–69
queen of the night (*Epiphyllum oxypetalum*) **246–47**
Quercus sp. (oak) 82–83, 133, 178
    Q. cerris (Turkey oak) 137
    Q. frainetto (Hungarian oak) 72
    Q. hartwissiana (strandzha oak) 307
    Q. macranthera (caucasian oak) 137
    Q. palustris (pin oak) 137
    Q. petraea (sessile oak) 137
    Q. suber (cork oak) 66
quiver tree (*Aloidendron dichotomum*) **146–47**

R

raceme inflorescences 216, 298
radicles (seeds) 22, *286*, 319
Rafflesia arnoldii (corpse flower) **260–61**
rainforest habitats
    cacti in *103*, 247
    epiphytes in 43
    flowers in 76, 211, 261
    leaves in 127, 135, 150, 168
    trees in 24, 80
    see also forest habitats
Ramonda sp. *235*
Ranunculaceae family (buttercup) 328
Ranunculus acris (buttercup) 217
raspberry (*Rubus idaeus*) 298–99
Ravenala madagascariensis (traveller's palm) 242, 309
ray flowers **218–19**, 221
red bartsia (*Odontites vulgaris*) 44
red buckeye (*Aesculus pavia*) 115
red clover (*Trifolium pratense*) 32–33
red colours
    beetroot 22
    bracts 278, 279
    flowers 248
    leaves 151, 153, 154
    see also anthocyanins
red mulberry (*Morus rubra*) 137
Redouté, Pierre-Joseph 270, 271
redwood, giant (*Sequoiadendron giganteum*) 111, 288
regal lily (*Lilium regale*) 324–35
regma 300, 326
Renaissance artwork **62–63**
reniform leaves 133
reproduction **174–75**, *233*, **282–83**
    see also clones; female parts, of flowers; male parts, of flowers
reptiles 227

resins **82–83**, 84
Rhaphidophora elliptifolia 97
Rhinanthus minor (yellow rattle) 44
rhizobia 33
rhizomes 86, *87*, 88, *90*, *104*
    aquatic plants *48*, 51
    bamboo 106
    contractile roots 36
Rhizophora sp. (mangrove) **54–55**
rhizosphere *48*
Rhodophiala bifida 301
rhomboid leaves 132
rhubarb, giant (*Gunnera manicata*) 131
ribbonwood (*Hoheria glabrata*) 133
Ribes sanguineum 198
rice 268
Ricinus communis (castor oil plant) 309
ridged bark 66
rings, tree 60, **64–65**, 90
Roman gardens 296
roots
    absorbing and storing nutrients 27,
        28–29, 32–33
    aerial **38–39**
    contractile **36–37**
    fibrous **20–21**
    in mangroves 52–53, 54
    in postimpressionist art **30–31**
    in water **48–49**, 51, 52–53, 54
    nodules 33
    of epiphytes and parasites 34, 41,
        42–43, 44, 47
    root systems **24–25**
    see also clones
    runners 104–05
    tap roots **22–23**
Rosa sp. (rose) 15, 79, 300
    R. canina (dog rose) 188, 193, 270,
        274–75
rosate flowers 188
rose *see Rosa* sp. (rose)
Roses on a Wall, Lambdin 281
rotate flowers 188
Rothschild's slipper orchid
        (*Paphiopedilum rothschildianum*)
        266–67
rowan tree (*Sorbus aucuparia*) 306–07
Roxburgh, William 92
Royal Botanic Gardens, Kew 271, 311,
        334
Rubus idaeus 301
Rubus sp. (blackberry) 298–99
Ruellia chartacea (wild petunia) 279
runners 104
Ruscus aculeatus (butcher's-broom) 60–61

S

Saccharum officinarum 'Ko-hapai' (sugar
        cane) 58
sacred lotus (*Nelumbo nucifera*) 144–45,
        332–33

Sagittaria sagittifolia (arrowhead) 133
sagittate leaves 133
saguaro cactus (*Carnegiea gigantea*)
        98–99, **100–01**
salt water plants 53, 54
    Rhizophora sp. (mangrove) **54–55**
samaras 153, 300, **314–15**
sap
    leaf 137, 155
    poisonous 103, 180, 231, 322
sapwood *64*
Sarraceniaceae family (carnivorous
        pitcher plants) 170, 256–57
saucer magnolia (*Magnolia* × *soulangeana*
        'Rustica Rubra') 308–09
Scabiosa stellata (scabious) 318–19
scales
    bark 66
    bracts 178, 231, *283*
    buds 73, 75, 118
    cones 283, 288, 290, 291
    leaves 43, 86, 111, *181*, 339
scars 72, 75, 181, 244
scents
    as deception 170, 211, 265
    for pollination 212, 227, 247, 255,
        258–59, 272
schizocarps 301, 324, 327
sclereids *167*
scorpioid cyme inflorescences 217
sea oats (*Uniola paniculata*) 223
seasons (flowering plants) **192–93**
secretory glands 265
seeds **286–87**, **308–09**,
    coats 333
    dispersal *277*, **304–05**, 317,
        320–21
        by animals 41, **306–07**,
        312–13, 247
        by explosion **326–27**
        by fire **330–31**
        by herbivores
        by water **332–33**
        by wind **314–15**, **318–19**, 322
        see also samaras
    enclosed **292–93**
    naked **288–89**, **290–91**
    see also cones
    of bananas 302
    of Japanese maples 153
    of sunflowers 221
    pods and capsules **324–25**
    see also fruits
segmented fruits 300
self-pollination 51, 80, 196, 201, 214,
    268–69
    see also cross-pollination
Sellars, Pandora 124
Selliguea plantaginea 338
sepals *184*, 185, 275, 276
    bird of paradise plant 262
    bleeding heart 269
    Chinese lantern 293

sepals *continued*
    lobster claw plant 279
    magnolias 74, 186
    orchids 258, 259, 265, 266
    pitcher plants 256
    shaving brush tree 245
septum 286, 287
Sequoiadendron giganteum (giant redwood)
    111, 288
serrate leaves 136
serrulate leaves 136
sessile oak (*Quercus petraea*) 137
Siebold, Philipp Franz von 205
silicles 286
silique fruits 301
silk 322
silky hairs 43, 75, 138, **320–21**
    see also trichomes
silver banksia (*Banksia marginata*) 242–43
silver birch (*Betula pendula*) 59, 333
silver leaves **148–49**, 165
silver-leafed mountain gum (*Eucalyptus
    pulverulenta*) 148–49
simple fruit *295*
sinuate leaves 137
slipper orchid (*Paphiopedilum*) 266–67
Smilax macrophylla *181*
smooth bark 67
snowdrop, giant (*Galanthus elwesii*) *238*
snowflakes, spring (*Leucojum vernum*) *254*
soil 27, *48*
    see also nutrients
Solanium sp. *235*, 300
solitary inflorescences 216
sonar *244*
sonication 235
Sorbus aucuparia (rowan tree) 306–07
sori 120, 336, 338, 340
South African thistle (*Berkheya purpurea*)
    160–61
spadix inflorescences 217
Spaendonck, Gerard van 271
Spanish moss (*Tillandsia* sp.) *43*
spathes 178, 209, 211, 263
spathulate leaves 133
species (classification) 16
spider plant (*Chlorophytum comosum*) 104
spike inflorescences 36, 85, 146, 217,
    242, 302
spikelet 222, 304, 305
spikes
    on scales (umbos) 290
    on stems 103
    see also prickles; spines; thorns
spines **78–79**, 158
    bract 176
    flower 276, *277*
    in cacti *98*, 99, 100, *102*, 247
    leaf 70, 160–61, **180–81**
    on seed capsules 312, 313
    pollen grains 198
spinous leaves 137
spinous stipules *181*

spiny bark 66
sporangia 338, 339, 340
spores
    ferns 120, 336, **338–39**, **340–41**
    fungi 34
    moss 343
sporophyte *340*, 341, 343
spring cherry (*Prunus* × *subhirtella*) 136
spring snowflakes (*Leucojum vernum*) *254*
squirrels 307
St. John's wort (*Hypericum pseudohenryi*)
    196–97
Stachys sp. 138–39, 217
stamens 77, *184*, 185, 197
    after fertilization 245, 298
    in magnolia 74
    in water lilies 51
staminal columns 201
star anise (*Illicium anisatum*) *186*
stems **58–59**, **60–61**
    "ant plants" 88–89
    cacti *98*, 100
    cauliflory **76–77**
    climbing plants **94–95**, **96–97**
    defences **78–79**, 80
    leafless **102–03**
    on aquatic plants 167
    resin **82–83**, 84
    runners **104–05**, 106
    trunks **64–65**, **66–67**, **90–91**
    see also branches; bulbs; corms;
    rhizomes
sterile male parts (flowers) *208*
stigma *184*, 185
    after fertilization 191, 197
    and pollination 51, 203, 214, 245,
        256, 261, 266, 268
        in grass plants *224*
        pin and thrum flowers *200*
stilt root systems *24*
    see also mangroves
stinging leaves **156–57**
stipules 158, 160, 180, 181
stolons 104
stomata 112, *126*, 129, 142, *148*
stork's bill (*Erodium cicutarium*) 300, *326*
storms 41, 53, 162
strangler fig (*Ficus* sp.) 40–41
strawberry, edible (*Fragaria* × *ananassa*)
    104, 300
strawberry, mock (*Duchesnea indica*, syn.
    Fragaria indica) 105
strawberry tree (*Arbutus unedo*) 294–295
Strelitzia reginae (bird of paradise plant)
    262–63, 309
striate bark 66
style *184*, 185, *196*, *256*, 299, 328
subspecies (classification) 16
succulent plants **102–03**, **142–43**, 146
    fruits 300–01
    stems *160*, *181*
    water storage in 98–99
    see also cacti

suckers *97*
sucrose *129*, 227, 240
sugar cane (*Saccharum officinarum* 'Ko-hapai') 58
sugar maple (*Acer saccharum* subsp. *saccharum*) 110–11
sugars
    by photosynthesis 34, 128, *129*
    in roots 22, 28
    in nectar 227
sundew (*Drosera* sp.) 170, **172–73**
sunflower (*Helianthus* sp.) **220–21**
sunlight *118*, **128–29**, 193, *272*
surimono, Hokusai 205
sweet chestnut *293*
sweet thorn (*Vachellia karroo*) **158–59**
Swiss cheese plant (*Monstera deliciosa*) 133, *134*
sycamore (*Acer pseudoplatanus*) 118–19, 314
symbiotic relationships *12*, 34, 88–89, **262–63**
symmetry, in flowers **188**
synconia 41, *263*
Syringa sp. (lilac) 216

T

Tamarindus indica (tamarind) 115
tap roots 20, *21*, **22–23**
Taraxacum sp. (dandelion) 23, 217, 218, **316–17**, 300, *321*
Tasman flax-lily (*Dianella tasmanica*) 306–07
Tasmanian snow tree (*Eucalyptus coccifera*) *148*
Taxus sp. (yew) 289
teasel, common (*Dipsacus fullonum*) 276–77
Tectaria pica 338
temperate habitats 51, 76, 106, 153, 316
temperature
    changes in 68, 153, 154, *272*, 273
    extremes 36, 47, 64, 146, 239
    see also weather extremes
tendrillar stipules *181*
tendrils *94*, 95, 96, *97*, 117, 171
tepals 74, 186
terminal buds 72, 196, 268
tetrads 341
thale cress (*Arabidopsis thaliana*) 133
Theobroma cacao (cocoa plant) 76
thigmotropism 95
thistles 160–61, 218, 231
thorns **78–79**, 158, 298
thrum flowers *200*
thyrse inflorescences 216
Tilia × *europaea* (common lime) 73
Tillandsia sp. 42–43
titan arum (*Amorphophallus titanum*) 210–11

toad lily (*Tricyrtis hirta*) 250–51
toddy palm (*Borassus flabellifer*) 92
toxic plants *see* poisonous plants
Trachycarpus fortunei (windmill palm) 133
Trachyspermum ammi 301
transpiration *126*, *142*
traps 158, 172, 176
    orchids *259*, 266
    pitcher plants 170–7, 257
    water lilies *212*
traveller's palm (*Ravenala madagascariensis*) 242, 309
tree fern (*Dicksonia antarctica*) *90*
tree frogs 168
tree of heaven (*Ailanthus altissima*) 136, 314
tree rings 60, **64–65**, 90
Tree Roots, van Gogh 30–31
tree trunks **64–65**, 83, **90–91**, 162
trichomes 138–39, 172, 229, 234
    bud defences 274, *275*
    silver leaves 148
    stinging leaves 156–57
Tricyrtis hirta (toad lily) 250–51
trifoliate leaves 115
Trifolium sp. (clover) 32–33, 115
tripinnate leaves 114
Tropaeolum majus (nasturtium) 133
tubers *87*, *104*
tubular flowers 189
Tulipa sp. (tulip) 216, *272*
Turkey oak (*Quercus cerris*) 137
Turkish hazel (*Corylus colurna*) 222
twining stems **94–95**
Typha latifolia (reed mace) 132

U

Ulmus glabra 300
ultraviolet (UV) light 250, 253, 264, 302
umbels 28, 214, 216
undulate leaves 137
unicorn plant (*Ibicella lutea*) *312*
unifoliate leaves 115
Uniola paniculata (sea oats) 223
unipinnate leaves 115
unisexual plants **206–07**
unscaled buds 72
unsegmented fruits 300
urceolate flowers 188
Urtica dioica (common nettle) 156–57

V

Vachellia karroo (sweet thorn) **158–59**
vacuoles *143*, *155*
valve (seeds) 287
van Gogh, Vincent 30–31
variegated leaves **150–51**
varieties (classification) 16

vascular cambium *64*
vascular tissues 14, 60–61, 163, 343
    parasitic plants 47, 261
    spines and thorns 79, 160, 161,
    see also veins (leaf); xylem
vegetative reproduction *174*
veins (leaf) 112–13, 130, 134–35, 150, 156
    aquatic plants 167
velamen *39*
velvet plant (*Gynura aurantiaca*) 138
venus fly traps 170
verticillaster inflorescences 217
vibrations, for pollination 234
Victoria amazonica (giant water lily) 166–67, 212–13
Vienna Dioscorides 141
Vinca sp. 198
Vinci, Leonardo da 63
vines 94–95, 97
Viola sp. (violet) 327
    V. reichenbachiana (early dog-violet) 136
    V. riviniana (common dog violet) 269
Virginia creeper (*Parthenocissus* sp.) 97
virgin's bower (*Clematis flammula*) 60–61
Viscum album (mistletoe) 44, **46–47**

W

Walcott, Mary Vaux 280–81
walnut (*Juglans* sp.) 72, 132
Warhol, Andy 236
Warren, Mary Shaffer 281
wasps *228*, 248, 249
    fig wasps 41, *263*
water **168–69**
    absorption *21*, 34, 38, *39*, 43
    see also succulent plants
    cycle (in leaves) **126–27**
    dispersal of seeds by **332–33**
    storage **98–99**, 100, 142, 143, 146
    see also aquatic plants
water lilies
    Nymphaea sp. 15, 16, **50–51**, 133
    Victoria amazonica (giant water lily) 166–67, 212–13
water-resistant plants *see* waxy surfaces
waxy surfaces
    carnivorous plants 170, *171*
    leaves 112, 126, *134*, **144–45**, 168
    seeds and fruits 47, 322
    silver leafed plants 148–49, 165
    succulent plants 98, 142
    see also suberin
weather extremes 131, **162–63**, 165
    droughts 37, 64, 98, 138, 146, 158
    storms 41, 53, 162
Weinmann, Johann Wilhelm 311
wetland plants 333
white coloured flowers 248
white leadtree (*Leucaena leucocephala*) 115

white water lily (*Nymphaea alba*) 133
whorled leaf arrangements *122*
whorled branches *70*
wild carrot (*Daucus carota*) 28–29, 216
wild cherry (*Prunus avium* 'Plena') 73
wild garlic (*Allium ursinum*) 216
wild petunia (*Ruellia chartacea*) 279
wildfires 158, 165, 331
    see also fire, release of seeds by
wildlife *see* animals
wind
    for seed spread 29, 43, 176, 304
        parachutes 317, 319, 321, 322
        samaras 314, 315
    for spore dispersal 338, 340, 341
    pollination by 71, 153, 317, 319, 321
        cones 291
        grasses **222–23**, 224, 225
    tolerant plants 162–63
windmill palm (*Trachycarpus fortunei*) 133
winged seeds (samaras) 153, 300, **314–15**
winter buds 72–73
winter flowering plants **238–39**
Wisteria 95
witch hazel (*Hamamelis* sp.) 74, 238–39
woodblock art 141, 205
woodland habitats 34
woody perennial plants 191
woody tree trunks 90

X

xylem *27*
    in aerial roots *39*
    in leaves 112, *113*, *126*
    in stems 59, 60, *64*, 98

Y

yellow pigments (carotenoids) 154, 155, 248
yellow rattle (*Rhinanthus minor*) 44
yew (*Taxus* sp.) 289
Yuan Jin 195

Z

Zea mays (corn) 301
Zingiber spectabile (beehive ginger) 279
Ziziphus mauritiana *181*
zygomorphic flowers 188

list of botanical art

16 *Sexual parts of flowers* Illustration by G.D. Ehret, from C. Linnaeus' *Genera Plantarum*, 1736

23 *Leontodon taraxacum* (syn. *Taraxacum officinale*) Copper engraving by W. Kilburn from W. Curtis's *Flora Londinensis*, London, 1777–98

30–31 *Tree Roots* Oil on canvas painting by V. van Gogh, 1890

31 *Cypress Trees* Polychrome and gold leaf screen by Kanō Eitoku, *c*.1890

33 *Trifolium subterraneum* Engraving by J. Sowerby from *English botany, or coloured figures of British plants*, 1864

43 *Tillandsia ionantha* Engraving from L. van Houtte, *Flore des serres et des jardin de l'Europe*, 1855

45 *Lathraea clandestina* Chromolithograph from *Revue horticole*, 1893

62 *Das große Rasenstück (Great Piece of Turf)* Watercolour by A. Dürer, 1503

63 *Two Trees on the Bank of a Stream* Drawing by L. da Vinci, *c*.1511–13

63 *Star of Bethlehem and other plants* Drawing by L. da Vinci, *c*.1505–10

74 *Forsythia viridissima* Drawing by W.H. Fitch from *Curtis's Botanical Magazine*, 1851

76 *Caracas theobroma* (syn. *Theobroma cacao*) Engraving from E. Denisse, *Flore d'Amérique*, 1843–46

89 *Myrmecodia beccari* Drawing by M. Smith from *Curtis's Botanical Magazine*, 1886

92 *Caesalpinia sappan* Hand painted copy of an illustration commissioned by W. Roxburgh.

92 *Borassus flabellifer* Drawing from W. Roxburgh's *Plants of the Coast of Coromandel*, 1795

93 *Livistona mauritiana* Watercolour by an unknown artist, part of Kew Collection

103 *Various Cactaceae* Engraving from Brockhaus' *Konversations-Lexikon*, 1796–1808

104–105 *Duchesnea indica* (syn. *Fragaria indica*) Drawing by Rungia from R. Wight, *Spicilegium Neilgherrense*, 1846

116–117 *Acanthus* wallpaper by W. Morris, 1875

117 Decorative stained-glass window by G. Beltrami, 1906

117 *Horse Chestnut leaves and blossom* Decorative study by J. Foord, 1901

120 *Aspidium filix-mas* (syn. *Dryopteris filix-mas*) Drawing by W. Müller from F.E. Köhler, *Medizinal-Pflanzen*, 1887

124 *Indian Onion painted in Benares 1971* Watercolour on vellum by R. McEwen, 1971

124 *Laelia tenebrosa, Philodendron hybrid, Calathea ornata, Philodendron leichtlinii, Polypodiaeceae* Watercolour by P. Sellars, 1989

125 *Brittany 1979* Watercolour on vellum by R. McEwen, 1979

138 *Mentha spicata* (syn. *Mentha viridis*) Engraving from G.C. Oeder *et al.*, *Flora Danica*, 1876

140 *Rubus sÿlvestris* Drawing from *Dioscorides Codex Vindobonensis Medicus Graecus*, 1460

141 *Chrysanthemum segetum* (left), *Leucanthemum vulgare* (right) Engraving by an unknown artist from Culpeper's *The English Physitian*, 1652

141 *Campanula* Engraving by an unknown artist from Culpeper's *The English Physitian*, 1652

176 *Passiflora foetida* (syn. *Passiflora gossypiifolia*) Drawing by S.A. Drake from *Edwards's Botanical Register*, 1835

186 *Illicium* Engraving from *Lehrbuch der praktischen Pflanzenkunde in Wort und Bild*, 1886

190 *Paeonia peregrina* Drawing by M. Smith from *Curtis's Botanical Magazine*, 1918

194–195 *Magnolia blossoms* Painting by C. Hongshou, from *Album of Fruit and Flowers*, 1540

195 *Blossoms* Painting by C. Chun

195 *Bird and flower painting* Painting by J. Jing, 1857

204 *Papaver* Colour woodblock print by I. Tsunemasa from *Honzo Zufu*, 1828

205 *Chrysanthemums* Colour woodblock print by K. Hokusai, 1831–33

205 *Mount Fuji with Cherry Trees in Bloom* Colour woodblock print by K. Hokusai, 1805

208 *Dieffenbachia seguine* Engraving from B. Maund, J.S. Henslow, *The botanist*, 1839

236 *Flowers* Print by A. Warhol, 1967

236–237 *Jimson weed* Painting by G. O'Keeffe, 1936

238 *Galanthus elwesii* Drawing by W.H. Fitch from *Curtis's Botanical Magazine*, 1875

242 *Red ruffed lemur* (*Varecia rubra*), *and ring-tailed lemur* (*Lemur catta*) Drawing by E. Travies from Charles d'Orbigny's *Dictionnaire Universel d'Histoire Naturelle*, 1849

257 *Sarracenia drummondii* Drawing from W. Robinson, *The garden. An illustrated weekly journal of horticulture in all its branches*, 1886

270 *Rosa centifolia* Stipple engraving, hand-finished in watercolour by P.J. Redouté from *Les Roses*, *c*.1824

271 *Narcissus × odorus* Watercolour by F. Bauer, *c*.1800

271 *Jacinthe Double* Painting by G. van Spaendonck, *c*.1800

280 *Dutchman's Pipe* Watercolour by M. Vaux Walcott from *North American Pitcher Plants*, 1935

281 *Roses on a Wall* Painting by G.D. Lambdin, 1877

296 Mosaic of a pomegranate tree from the Great Palace of Constantinople, (*c*.575–641)

296–297 Fresco from Villa of Livia

310 *Malus* Mezzotint engraving, coloured by hand, G.D. Ehret from *Phytanthoza Iconographia*, 1737–45

311 *Ananas sativus* Drawing by G.D. Ehret

312 *Ibicella lutea* Drawing and engraving by F. Burmeister from *Flora Argentina*, 1898

334 *Carica papaya* Chromolithograph from B. Hoola van Nooten's *Fleurs, Fruits et Feuillages*, 1863–64

334 *Myristica fragrans* Oil on canvas painting by M. North, 1871–72

335 *Blighia sapida* Oil on canvas painting by M. North

acknowledgments

The Publisher would like to thank the directors and staff at the Royal Botanic Gardens, Kew for their enthusiastic help and support throughout the preparation of this book, in particular Richard Barley, Director of Horticulture; Tony Sweeney, Director of Wakehurst; and Kathy Willis, Director of Science. Special thanks to all at Kew Publishing, especially Gina Fullerlove, Lydia White, and Pei Chu, and to Martyn Rix for his detailed comments on the text. Thanks to the Kew Library, Art, and Archives team, particularly Craig Brough, Julia Buckley, and Lynn Parker, and also to Sam McEwen and Shirley Sherwood.

DK would also like to thank the many people who provided help and support with photoshoots in the tropical nursery and gardens at Kew and Wakehurst, and all who provided expert advice on specific details, notably Bill Baker, Sarah Bell, Mark Chase, Maarten Christenhusz, Chris Clennett, Mike Fay, Tony Hall, Ed Ikin, Lara Jewett, Nick Johnson, Tony Kirkham, Bala Kompalli, Carlos Magdalena, Keith Manger, Hugh McAllister, Kevin McGinn, Greg Redwood, Marcelo Sellaro, David Simpson, Raymond Townsend, Richard Wilford, and Martin Xanthos.

The Publisher would also like to thank Sylvia Myers and her team of volunteers at the London Wildlife Trust's Centre for Wildlife Gardening (www.wildlondon.org), and Rachel Siegfried at Green and Georgeous flower farm in Oxfordshire for hosting photogaphic shoots. DK is also grateful to Joannah Shaw of Pink Pansy and Mark Welford of Bloomsbury Flowers for their help in sourcing plants for photoshoots, and to Dr Ken Thompson for his help in the early stages of this book.

DK would also like to thank the following:

Additional picture research:
Deepak Negi

Image retoucher: Steve Crozier

Creative Technical Support:
Sonia Charbonnier, Tom Morse

Proofreader: Joanna Weeks

Indexer: Elizabeth Wise

PICTURE CREDITS

The publisher would like to thank the following for their kind permission to reproduce their photographs:

(Key: a-above; b-below/bottom; c-centre; f-far; l-left; r-right; t-top)

4–5 Alamy Stock Photo: Gdns81 / Stockimo.
6 Dorling Kindersley: Green and Gorgeous Flower Farm.
8–9 500px: Azim Khan Ronnie.
10–11 iStockphoto.com: Grafissimo.
12 Dorling Kindersley: Neil Fletcher (tr); Mike Sutcliffe (tc).
14 Alamy Stock Photo: Don Johnston PL (fcr). **Getty Images:** J&L Images / Photographer's Choice (c); Daniel Vega / age fotostock (cr). **Science Photo Library:** BJORN SVENSSON (cl).
15 Dorling Kindersley: Gary Ombler: Centre for Wildlife Gardening / London Wildlife Trust (br). **FLPA:** Arjan Troost, Buiten-beeld / Minden Pictures (c). **iStockphoto.com:** Alkalyne (cl).
16 Alamy Stock Photo: Mark Zytynski (cb).
17 Alamy Stock Photo: Pictorial Press Ltd.
18–19 iStockphoto.com: ilbusca.
20 Science Photo Library: Dr. Keith Wheeler (clb).
20–21 iStockphoto.com: Brainmaster.
22–23 iStockphoto.com: Pjohnson1.
23 Alamy Stock Photo: Granger Historical Picture Archive (br).
24–25 Getty Images: Ippei Naoi.
26 Alamy Stock Photo: Emmanuel Lattes.
30–31 Alamy Stock Photo: The Protected Art Archive (t).
31 Alamy Stock Photo: ART Collection (br).
32–33 Science Photo Library: Gustoimages.
33 Getty Images: Universal History Archive / UIG (tr). **Science Photo Library:** Dr. Jeremy Burgess (br).
34–35 Amanita: facebook.com/ Amanlta/ 500px.com/sot1s.
40–41 Thomas Zeller / Filmgut.
41 123RF.com: Mohammed Anwarul Kabir Choudhury (b).
43 © Board of Trustees of the Royal Botanic Gardens, Kew: (br).
44 123RF.com: Richard Griffin (bc, br). **Dreamstime.com:** Kazakovmaksim (l).
45 © Board of Trustees of the Royal Botanic Gardens, Kew.
46–47 Dreamstime.com: Rootstocks.
47 Alamy Stock Photo: Alfio Scisetti (tc).
50–51 Getty Images: Michel Loup / Biosphoto.
52–53 Rosalie Scanlon Photography and Art, Cape Coral, FL., USA.
53 Alamy Stock Photo: National Geographic Creative (br).
54 Alamy Stock Photo: Angie Prowse (bl).

54–55 Alamy Stock Photo: Ethan Daniels.
56–57 iStockphoto.com: Nastasic.
58 Getty Images: Davies and Starr (clb). **iStockphoto.com:** Ranasu (r).
59 iStockphoto.com: Wabeno (r).
60–61 Science Photo Library: Dr. Keith Wheeler (b); Edward Kinsman (t).
61 Science Photo Library: Dr. Keith Wheeler (crb); Edward Kinsman (cra).
62 Alamy Stock Photo: Interfoto / Fine Arts.
63 Alamy Stock Photo: The Print Collector / Heritage Image Partnership Ltd (cl, clb).
64–65 Getty Images: Doug Wilson.
66 123RF.com: Nick Velichko (c). **Alamy Stock Photo:** blickwinkel (tc); Joe Blossom (tr). **iStockphoto.com:** Westhoff (br).
67 Dorling Kindersley: Mark Winwood / RHS Wisley (bl). © **Mary Jo Hoffman:** (r).
68 Alamy Stock Photo: Christina Rollo (bl).
68–69 © Aaron Reed Photography, LLC.
71 Getty Images: Nichola Sarah (tr).
74 © Board of Trustees of the Royal Botanic Gardens, Kew: (bl).
76 Denisse, E., Flore d'Amérique, t. 82 (1843–46): (bc).
80–81 Getty Images: Gretchen Krupa / FOAP.
82–83 iStockphoto.com: Cineuno.
83 Science Photo Library: Michael Abbey (bc).
84–85 Don Whitebread Photography.
85 Alamy Stock Photo: Jurate Buiviene (cb).
86–87 Getty Images: Peter Dazeley / The Image Ban.
88 FLPA: Mark Moffett / Minden Pictures (bl).
89 © Board of Trustees of the Royal Botanic Gardens, Kew.
90–91 Alamy Stock Photo: Juergen Ritterbach.
92 Bridgeman Images: Borassus flabelliformis (Palmaira tree) illustration from *Plants of the Coromandel Coast*, 1795 (coloured engraving), Roxburgh, William (fl.1795) / Private Collection / Photo © Bonhams, London, UK (tl). © **Board of Trustees of the Royal Botanic Gardens, Kew:** (crb).
93 © Board of Trustees of the Royal Botanic Gardens, Kew.
94–95 Alamy Stock Photo: Arco Images GmbH.
95 Alamy Stock Photo: Alex Ramsay (br).
100 123RF.com: Curiousotter (bl).
100–101 © Colin Stouffer Photography.
103 Alamy Stock Photo: FL Historical M (tr).
104–105 © Board of Trustees of the Royal Botanic Gardens, Kew. 106–107 Ryusuke Komori.

106 © Mary Jo Hoffman: (bc).
108–109 iStockphoto.com: Engin Korkmaz.
111 Alamy Stock Photo: Avalon / Photoshot License (tr).
112 Science Photo Library: Eye of Science (cl).
112–113 Damien Walmsley.
115 123RF.com: Amnuay Jamsri (tl); Mark Wiens (tc). **Dreamstime.com:** Anna Kucherova (cr); Somkid Manowong (tr); Phanuwatn (cl); Poopiaw345 (bl); Yekophotostudio (br).
116–117 Alamy Stock Photo: Granger, NYC. / Granger Historical Picture Archive.
117 Alamy Stock Photo: Chronicle (cb). **Getty Images:** DEA / G. Cigolini / De Agostini Picture Library (tc).
118 Alamy Stock Photo: Richard Griffin (cr); Alfio Scisetti (cl).
119 Alamy Stock Photo: Richard Griffin.
120 © Board of Trustees of the Royal Botanic Gardens, Kew: (cla).
124 © Pandora Sellars / From the Shirley Sherwood Collection with kind permission: (tl). © **The Estate of Rory McEwen:** (crb).
125 © The Estate of Rory McEwen.
130–131 iStockphoto.com: lucentius.
131 123RF.com: gzaf (tc). **Alamy Stock Photo:** Christian Hütter / imageBROKER (tr). **iStockphoto.com:** lucentius (cla).
132 Alamy Stock Photo: Ron Rovtar / Ron Rovtar Photography (cl).
133 Alamy Stock Photo: Leonid Nyshko (cl); Bildagentur-online / Mc-Photo-BLW / Schroeer (tr). **Dorling Kindersley:** Gary Ombler: Centre for Wildlife Gardening / London Wildlife Trust (bl). **iStockphoto. com:** ChristineCBrooks (br); joakimbkk (cr).
134 FLPA: Ingo Arndt / Minden Pictures (br).
135 iStockphoto.com: Enviromantic (cr).
136 Dorling Kindersley: Batsford Garden Centre and Arboretum (bc). **Dreamstime. com:** Paulpaladin (cl).
137 Dorling Kindersley: Batsford Garden Centre and Arboretum (cl). **iStockphoto. com:** ByMPhotos (c); joakimbkk (cr).
138 Alamy Stock Photo: Val Duncan / Kenebec Images (c). **Dorling Kindersley:** Centre for Wildlife Gardening / London Wildlife Trust (tr). © **Board of Trustees of the Royal Botanic Gardens, Kew:** (cl).
138–139 Dorling Kindersley: Centre for Wildlife Gardening / London Wildlife Trust.
140 Bridgeman Images: Rubus sylvestris / Natural History Museum, London, UK.
141 Getty Images: Florilegius / SSPL (tl, cr).
143 Getty Images: Nigel Cattlin / Visuals Unlimited, Inc. (tl).
144 123RF.com: Sangsak Aeiddam.
145 Science Photo Library: Eye of Science (cra).
146 Alamy Stock Photo: Daniel Meissner (bl).

146–147 © Anette Mossbacher Landscape & Wildlife Photographer. 148 Science Photo Library: Eye Of Science (clb). **148–149 © Mary Jo Hoffman. 150 123RF.com:** Nadezhda Andriiakhina (bc). **Alamy Stock Photo:** allotment boy 1 (cl); Anjo Kan (cr). **iStockphoto.com:** NNehring (br). **Jenny Wilson / flickr. com/photos/jenthelibrarian/:** (c). **151 Dreamstime.com:** Sirichai Seelanan (bl). **152–153 © Aaron Reed Photography, LLC. 153 iStockphoto.com:** xie2001 (tr). **154 Science Photo Library:** John Durham (bl). **154–155 © Mary Jo Hoffman. 156 Science Photo Library:** Power and Syred (tc). **158 123RF.com:** Grobler du Preez (bl). **158–159** Dallas Reed. **162–163 iStockphoto.com:** DNY59. **164–165 © Josef Hoflehner. 165 Benoît Henry:** (br). **172–173 Getty Images:** Ralph DeseniĂŸ. **175 Dreamstime.com:** Arkadyr (cr). **176 © Board of Trustees of the Royal Botanic Gardens, Kew:** (cla). **178 123RF.com:** ncristian (bc); Nico Smit (c). **Alamy Stock Photo:** Zoonar GmbH (cl). **Dreamstime.com:** Indigolotos (br). **179 Getty Images:** Alex Bramwell / Moment. **182–183 iStockphoto.com:** Ilbusca. **186 Getty Images:** bauhaus1000 / DigitalVision Vectors (bc). **188 123RF.com:** godrick (cl); westhimal (tc); studio306 (bl). **Alamy Stock Photo:** Lynda Schemansky / age fotostock (br); Zoonar GmbH (cr). **Dorling Kindersley:** Gary Ombler: Centre for Wildlife Gardening / London Wildlife Trust (c). **189 123RF.com:** Anton Burakov (tl); Stephen Goodwin (tc, br); Boonchuay Iamsumang (tr); Oleksandr Kostiuchenko (cl); Pauliene Wessel (c); shihina (bc). **Getty Images:** joSon / Iconica (bl). **190 Alamy Stock Photo:** Paul Fearn (cla). **193 Dorling Kindersley:** Gary Ombler: Centre for Wildlife Gardening / London Wildlife Trust (bl). **194–195 Bridgeman Images:** Blossoms, one of twelve leaves inscribed with a poem from an Album of Fruit and Flowers (ink and colour on paper), Chen Hongshou (1768–1821) / Private Collection / Photo © Christie's Images. **195 Alamy Stock Photo:** Artokoloro Quint Lox Limited (cb). **Mary Evans Picture Library:** © Ashmolean Museum (tr). **198 Science Photo Library:** AMI Images (tl); Power and Syred (tc, tr); Steve Gschmeissner (cla, ca, cra, cl, c); Eye of Science (cr).

198–199 Alamy Stock Photo: Susan E. Degginger. **202–203 Dorling Kindersley:** Centre for Wildlife Gardening / London Wildlife Trust. **204 © Board of Trustees of the Royal Botanic Gardens, Kew. 205 Getty Images:** Katsushika Hokusai (cl). **Rijksmuseum Foundation, Amsterdam:** Gift of the Rijksmuseum Foundation (tr). **206 Alamy Stock Photo:** Wolstenholme Images. **207 Alamy Stock Photo:** Rex May (cla). **Getty Images:** Ron Evans / Photolibrary (cl). **208 © Board of Trustees of the Royal Botanic Gardens, Kew:** (br). **210–211 © Douglas Goldman, 2010. 211 Getty Images:** Karen Bleier / AFP (br). **212 Alamy Stock Photo:** Nature Picture Library (tl). **213 SuperStock:** Minden Pictures / Jan Vermeer. **214–215 Dorling Kindersley:** Gary Ombler: Centre for Wildlife Gardening / London Wildlife Trust. **216 123RF.com:** Aleksandr Volkov (bl). **Getty Images:** Margaret Rowe / Photolibrary (tr). **iStockphoto.com:** narcisa (cr); winarm (c); Zeffss1 (cl). **217 Alamy Stock Photo:** Tamara Kulikova (tl); Timo Viitanen (c). **Dreamstime.com:** Kazakovmaksim (br). **Getty Images:** Frank Krahmer / Photographer's Choice RF (tc). **iStockphoto.com:** AntiMartina (bc); cjaphoto (cr). **220–221 Dreamstime.com:** Es75. **221 Getty Images:** Sonia Hunt / Photolibrary (br). **223 FLPA:** Minden Pictures / Ingo Arndt (crb). **227 FLPA:** Minden Pictures / Mark Moffett (tl). **228 Getty Images:** Jacky Parker Photography / Moment Open (tl). **230–231 Getty Images:** Chris Hellier / Corbis Documentary. **233 Alamy Stock Photo:** ST-images (tc). **234–235 Dorling Kindersley:** Gary Ombler: Centre for Wildlife Gardening / London Wildlife Trust. **235 Alamy Stock Photo:** dpa picture alliance (tr). **236 Bridgeman Images:** Jimson Weed, 1936–37 (oil on linen), O'Keeffe, Georgia (1887–1986) / Indianapolis Museum of Art at Newfields, USA / Gift of Eli Lilly and Company / © Georgia O'Keeffe Museum / © DACS 2018. **237 Alamy Stock Photo:** Fine Art Images / Heritage Image Partnership Ltd / © 2018 The Andy Warhol Foundation for the Visual Arts, Inc. / Licensed by DACS, London. / © DACS 2018 (cb). **238 Alamy Stock Photo:** Neil Hardwick

(cl). **© Board of Trustees of the Royal Botanic Gardens, Kew:** (bc). **242 Getty Images:** Florilegius / SSPL (cla). **243 Alamy Stock Photo:** Dave Watts (cra). **244 FLPA:** Photo Researchers (bl). **244–245 Image courtesy Ronnie Yeo. 246–247 Getty Images:** I love Photo and Apple. / Moment. **247 Alamy Stock Photo:** Fir Mamat (tr). **250 Science Photo Library:** Cordelia Molloy (bc). **252 123RF.com:** Jean-Paul Chassenet (bc). **252–253 Craig P. Burrows. 254 Alamy Stock Photo:** WILDLIFE GmbH (tc). **257 © Board of Trustees of the Royal Botanic Gardens, Kew:** (br). **260–261 iStockphoto.com:** mazzzur (l). **261 Getty Images:** Paul Kennedy / Lonely Planet Images (tr). **263 Johanneke Kroesbergen-Kamps:** (tr). **265 Dreamstime.com:** Junko Barker (crb). **268–269 Alamy Stock Photo:** Alan Morgan Photography. **269 Alamy Stock Photo:** REDA &CO srl (br). **270 Bridgeman Images:** *Rosa centifolia*, Redouté, Pierre-Joseph (1759–1840) / Lindley Library, RHS, London, UK. **271 Alamy Stock Photo:** ART Collection (clb); The Natural History Museum (tr). **274–275 Dorling Kindersley:** Centre for Wildlife Gardening / London Wildlife Trust. **275 Dorling Kindersley:** Centre for Wildlife Gardening / London Wildlife Trust (tr). **280 Smithsonian American Art Museum:** Mary Vaux Walcott, White Dawnrose (*Pachyloplus marginatus*), n.d., watercolor on paper, Gift of the artist, 1970.355.724. **281 Bridgeman Images:** Roses on a Wall, 1877 (oil on canvas), Lambdin, George Cochran (1830–96) / Detroit Institute of Arts, USA / Founders Society purchase, Beatrice W. Rogers fund (cr). **284–285 iStockphoto.com:** Ilbusca. **289 123RF.com:** Valentina Razumova (br). **290–291 Dreamstime.com:** Tomas Pavlasek. **291 Dreamstime.com:** Reza Ebrahimi (crb). **292–293 Getty Images:** Mandy Disher Photography / Moment. **294 Gerald D. Carr:** (bl). **296 Alamy Stock Photo:** Picade LLC (tl). **296–297 Alamy Stock Photo:** Hercules Milas. **298–299 Dorling Kindersley:** Centre for Wildlife Gardening / London Wildlife Trust. **300 123RF.com:** Vadym Kurgak (cl). **Alamy Stock Photo:** Picture Partners (tl); WILDLIFE GmbH (c/Wych Elm).

iStockphoto.com: anna1311 (c/Achene); photomaru (tc/Hesperidium); csundahl (tc/Pepo). **301 123RF.com:** Dia Karanouh (c). **Alamy Stock Photo:** deefish (c/Legume); John Kellerman (tl); Jurij Kachkovskij (tc/Peach); Shullye Serhiy / Zoonar (tr). **Dorling Kindersley:** Centre for Wildlife Gardening / London Wildlife Trust (bc). **iStockphoto. com:** anna1311 (tc/Aggregate); ziprashantzi (br). **302 123RF.com:** Saiyood Srikamon (bl). **302–303 Julie Scott Photography. 304 Getty Images:** Ed Reschke / Photolibrary (cl). **310 Bridgeman Images:** Various apples (with blossom), 1737–45 (engraved and mezzotint plate, printed in colours and finished by), Ehret, Georg Dionysius (1710–70) (after) / Private Collection / Photo © Christie's Images. **311 Alamy Stock Photo:** The Natural History Museum. **312 © Board of Trustees of the Royal Botanic Gardens, Kew:** (cla). **313 Dreamstime.com:** Artem Podobedov / Kiorio (tl). **316–317 Jeonsango / Jeon Sang O. 319 Science Photo Library:** Steve Gschmeissner (br). **322–323 Getty Images:** Don Johnston / All Canada Photos. **322 © Mary Jo Hoffman:** (bc). **326–327 Alamy Stock Photo:** Blickwinkel. **327 Alamy Stock Photo:** Krystyna Szulecka (cr); Duncan Usher (tr). **328–329 © Michael Huber. 333 iStockphoto.com:** heibaihui (br). **334 Bridgeman Images:** Papaya: *Carica papaya*, from Berthe Hoola van Nooten's *Fleurs, Fruits et Feuillages* / Royal Botanical Gardens, Kew, London, UK (tr). **© Board of Trustees of the Royal Botanic Gardens, Kew:** (cl). **335 Science & Society Picture Library:** The Board of Trustees of the Royal Botanic Gardens, Kew. **339 Getty Images:** André De Kesel / Moment Open (tc). **340 Getty Images:** Ed Reschke / The Image Bank (tl). **342–343 Getty Images:** Peter Lilja / The Image Bank. **343 Alamy Stock Photo:** Zoonar GmbH (cb).

All other images © Dorling Kindersley
For further information see:
www.dkimages.com